Appreciation of Villa Styles

别墅豪宅风格赏析

佳图文化 编

·广州·

图书在版编目（CIP）数据

别墅豪宅风格赏析：汉英对照 / 佳图文化编. — 广州 ：华南理工大学出版社，2014.10
ISBN 978-7-5623-4400-1

Ⅰ. ①别… Ⅱ. ①佳… Ⅲ. ①别墅 – 室内装饰设计 – 世界 – 图集 Ⅳ. ① TU241-64

中国版本图书馆 CIP 数据核字（2014）第 211252 号

别墅豪宅风格赏析
佳图文化 编

出 版 人：韩中伟
出版发行：华南理工大学出版社
（广州五山华南理工大学 17 号楼，邮编 510640）
http://www.scutpress.com.cn E-mail: scutc13@scut.edu.cn
营销部电话：020-87113487 87111048（传真）
策划编辑：赖淑华
责任编辑：骆 婷 赖淑华
印 刷 者：利丰雅高印刷（深圳）有限公司
开 本：889mm×1194mm 1/12 印张：28
成品尺寸：285mm×285mm
版 次：2014 年 10 月第 1 版 2014 年 10 月第 1 次印刷
定 价：388.00 元

PREFACE 前言

Luxury villas represent people's dream for ideal habitat and ideal living.

"Appreciation of Villa Styles" is a professional book that fully introduces and presents excellent style designs for luxurious villas all around China. The selected villas and houses are designed in modern style, neoclassical style, luxurious European style, simple European style, mixed Chinese-Western style, new Chinese style, or Art Deco style, presenting different flavors and charms. All these projects are well introduced and analyzed to enable readers to experience the beauty of the residences and the unique ideas of the designers. With a large number of beautiful images, it will give the readers more intuitive cognition and visual experience of beautiful residential design. For the professional readers, it will be a great reference book for their practice and research on the design of villa styles.

别墅豪宅是人类对自己的理想居所与理想生活的极致追求。

《别墅豪宅风格赏析》是一本全面介绍和展示国内优秀别墅豪宅风格设计案例的专业书籍。本书精选国内极具代表性的别墅豪宅风格设计，涵盖现代风格、新古典风格、奢华欧式风格、简约欧式风格、中西混搭风格、新中式风格、Art Deco风格等。本书深入浅出，对每一个案例都进行了全面赏析，力求让读者感受到住宅设计之美，体会到设计师心思之巧。同时，本书配以大量精美实景图，希望能让读者有更为直观的认知与视觉体验，从而感受到住宅设计的美之真谛。对于实践以及研究别墅豪宅风格设计的专业人士来说这是一本重要的参考书籍。

CONTENTS

Neoclassical Style 新古典风格

Art Deco Style Art Deco风格

Luxurious European Style 奢华欧式风格

Simple European Style 简约欧式风格

目录

Mixed Chinese-western Style 中西混搭风格

New Chinese Style 新中式风格

Modern Style 现代风格

Other Styles 其他风格

Neoclassical Style

新古典风格

Beautiful and Harmonious
唯美和谐

Warm and Romantic
浪漫温馨

Elegant and Dignified
典雅厚重

Vantone Legacy Town
万通新新家园

Designer: Wang Xiaogen
Interior Design: Beijing Genshang International Space Design Co., Ltd.
Location: Shunyi District, Beijing, China
Area: 365 m^2

设计师：王小根
室内设计：北京根尚国际空间设计有限公司
项目地点：中国北京市顺义区
面积：365 m^2

Keywords 关键词

French Neoclassical 法式新古典

Modern Abstract Art 现代抽象艺术

Oriental Culture Elements 东方文化元素

Furnishings/Materials 软装 / 材料

White Mix Oil Paint Wood Veneer, Hand-painted Wallpaper, Copper Bar, Block Wood Floor, Carpet

白色混油木饰面、手绘壁纸、铜条、拼花木地板、地毯

This project is a three-storey house. The first floor is public area, the second floor is master bedroom area, the third floor is the elderly room and children room area, the underground is the recreational area, elevator is designed to make it easy to go up and down stairs. Combining the French neoclassical and modern abstract art, this project also has some oriental elements. The elegant French classic blue and yellow silk fabric matched with abstract carpet patterns, painting, lamps and lanterns, presenting a delicate, exquisite, free and bold space style. The utilization of some oriental elements adds some generous and solemn elements in luxury elements.

The whole space takes blue and yellow as the main tone, and takes black and white to neutralize the space color. The wall at the hallway takes the collocation of light color clapboard and distressed mirror. The front of the hallway is the kitchen, and the kitchen is enclosed by the breakfast table skillfully, being independent from the hallway and also connecting the hallway. The right dining room and living room use four white "Doric Order" to make the space more generous, while the white columns also blend into the space naturally.

The space pays attention to smooth lines and visual transparent feeling. The living room with a 7.5 meter width has two meeting function areas, making full use of the room. The two function areas use the same bag and single sofa to keep the unity meanwhile keep independent.

The master bedroom area is placed on the second floor and is divided into left and right coherent parts by the hallway. The right is bedroom area and reading area, and the left is walk-in closet and main bathroom. The whole space is transparent and the function layout is reasonable. The low purity of blue and yellow makes the space more soft and comfortable; the abstract print of carpet is the modern art technique of expression, while the use of blue makes the modern elements blended into the whole space harmoniously.

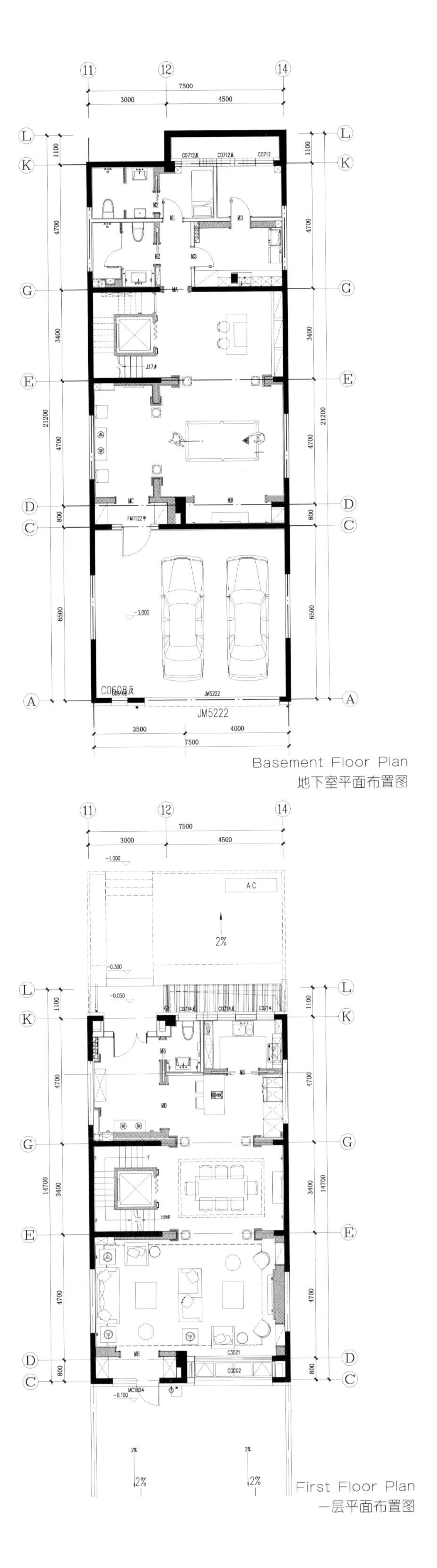

Basement Floor Plan
地下室平面布置图

First Floor Plan
一层平面布置图

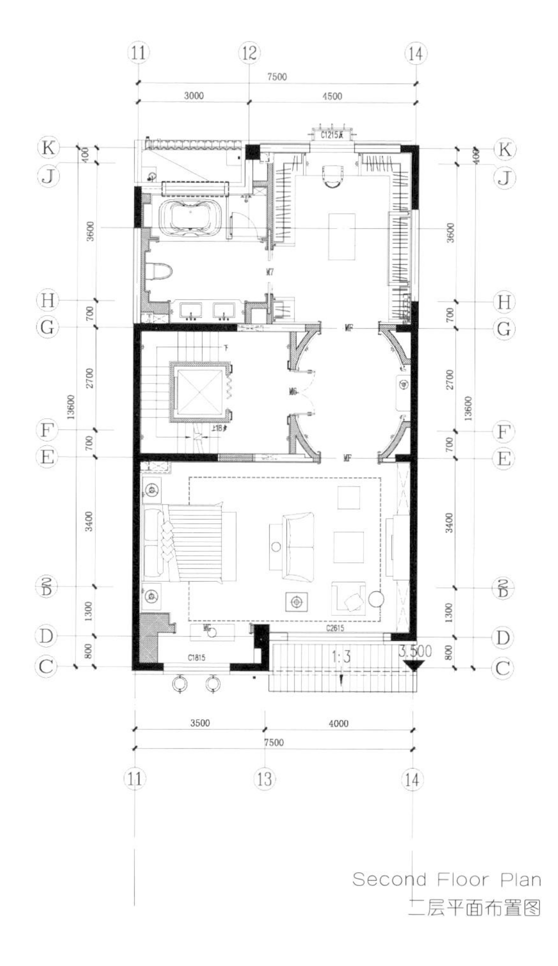

Second Floor Plan
二层平面布置图

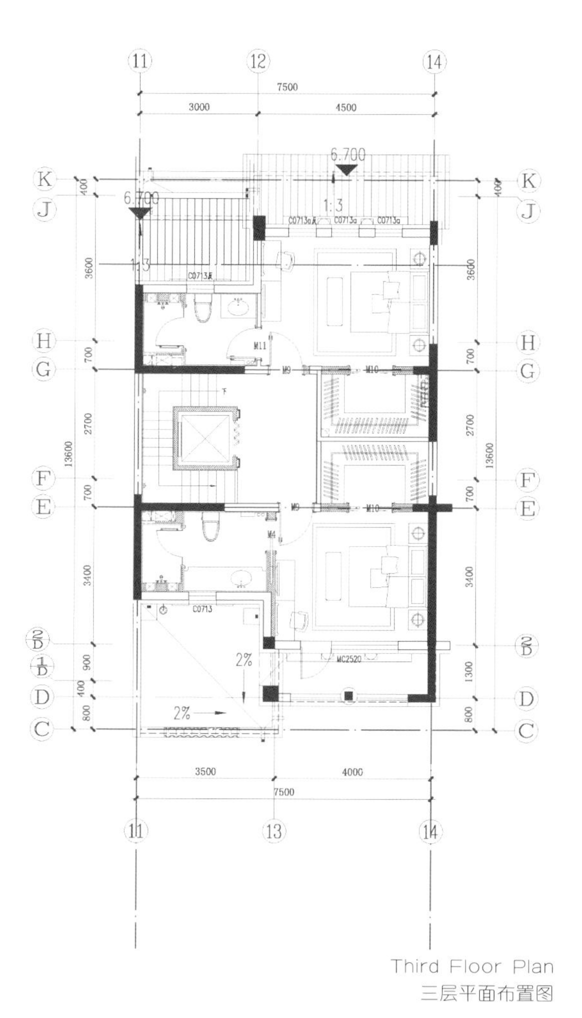

Third Floor Plan
三层平面布置图

项目户型为一栋上下三层与地下室：一层公共区、二层主卧区、三层老人房与儿童房，地下是休闲区，上下楼层设计了电梯来辅助行动。作为一个法式新古典格调与现代抽象艺术并重的别墅，项目同时伴有少量东方元素空间。优雅的法式经典蓝色、黄色丝质布艺与抽象地毯纹样、挂画、灯具相搭配，展现了精致、细腻兼具自由、奔放的空间格调。少量东方文化元素的运用，在奢华中增添了些许大气与庄重。

整体空间以蓝、黄色为主基调，使用中性的黑、白色中和空间色彩。入口玄关处的墙面使用浅色护墙板与做旧镜面的搭配。门厅的前方是厨房，早餐台将厨房巧妙地围合起来，与门厅既独立又相连成整体。右侧的餐

厅与客厅区域使用了4根白色“多里克式柱”使空间更加大气，白色柱体自然地融入空间。

空间注重顺畅的动线与视觉的通透感觉。横向7.5 m的客厅面积，设计有两个会客功能区，使利用率得到充分发挥。两组功能区利用相同的靠包、单人沙发，保持了空间的独立，同时统一为一个整体。

主卧区设置在二层，通过玄关将整个二层分为左右两个连贯的区域，右边为卧室区、阅读区，左边为步入式衣帽间和主卫间，使整个空间通透，功能布局安排合理。低纯度的蓝色、黄色的使用增加了空间的柔和与舒适。地毯的抽象图案是现代艺术的表现手法，而蓝色的使用使得现代元素和谐地融入整体空间。

FENDI

GZLakes Show Flat of 21#D

广州金地荔湖城 21#D 户型样板房

Designer: Lee Jianming
Interior Design: Dongguan Scale Design Co., Ltd.
Location: Zengcheng District, Guangzhou, Guangdong, China
Area: 547 m^2

设计师：李坚明
室内设计：东莞市尺度室内设计有限公司
项目地点：广东省广州市增城区
面积：547 m^2

Keywords 关键词

Neoclassical Style 新古典风格

Low-Key and Clam 低调沉稳

Sense of Belonging 归属感

Furnishings/Materials 软装 / 材料

Imported Marble, Emperor Gold, Wallpaper, Leather, Silver Mirrors, Rose Gold Stainless Steel

进口大理石、帝皇金、墙纸、皮革、银镜、玫瑰金不锈钢

In order to meet the requirements of the upstart class in Guangzhou, the project is oriented to be in neoclassical style. Adopting rich artistic deposits of the European culture, revealing the creative design concept from the furniture and soft furnishings, the project meets the aesthetic and cultural psychology requirements of the target residents, and expresses the new idea of comfortable lives and gives residents a sense of belonging and comfort.

The simple and clear lines outline the charm of the neoclassical style. Cooperating with proper decoration, they are explaining the quiet and elegant lives quite well, which have been the first choice of whom are pursuing for high quality lives. Selecting neoclassical style, low-key and clam color tone for this project, its space design is quite luxury and superb. The designers abandon complicated decorations, but use the most classical elements and simple methods to show the sense of history and culture of the project. Lots of decorative materials are used for the unit, such as silver foil wallpaper, stone, wood, tiles and so on, since they are in a unified style, it still reveals the concise atmosphere.

Basement Floor Plan
负一层平面布置图

项目为迎合广州的高端新贵阶层购房者的喜好，将设计风格定位为新古典风格。设计吸取欧洲文化中丰富的艺术底蕴，从家具、软装中体现开放创新的设计思想及项目的尊贵姿容，恰到好处地满足现代“新贵族”的审美和文化心理需求。温馨中体现了现代人对享受生活的新主张，让人置身其中有一种自然而然的归宿感和舒适感。

简洁、明晰的线条勾勒出欧洲新古典的神韵，配上得体有度的装饰，让人感受到恬静典雅、悠闲舒适，成为现代人追求高品质生活的首选。设计师用新古典风格诠释项目，低调沉稳的中性色调散发出无法掩盖的气势。整个空间设计大气奢华，无处不给人华丽的感受。设计表达的是具有厚度的形式美，摒弃繁复的艳媚装饰；以最经典的古典元素，最简约的表现手法，展现历史感和文化纵深感。项目虽然使用了很多材质如银箔壁纸、石材、木材、瓷砖等，但是在整体把控上严谨统一，使人们有一种舒适的感觉。

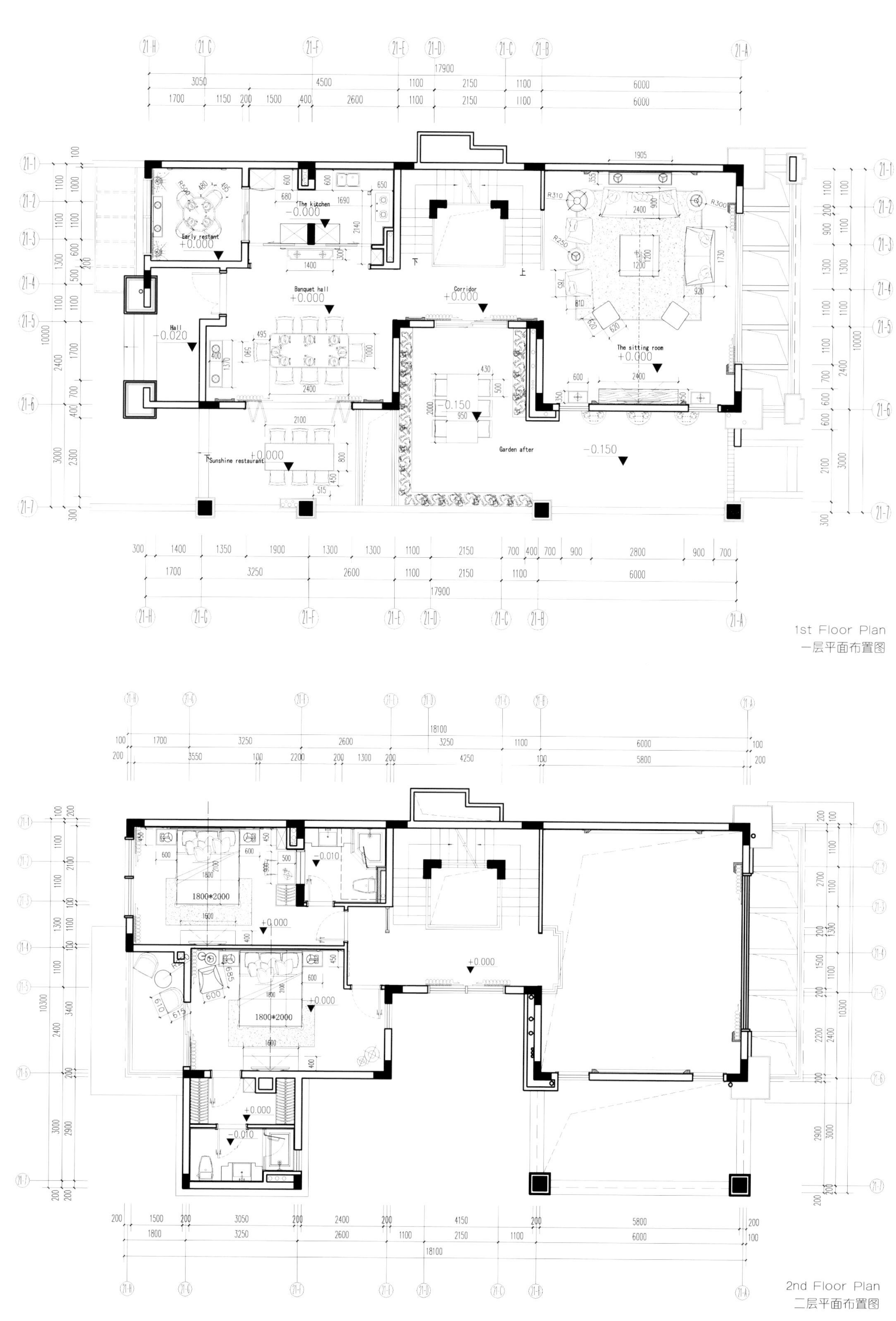

1st Floor Plan
一层平面布置图

2nd Floor Plan
二层平面布置图

BOOK
SOCKS

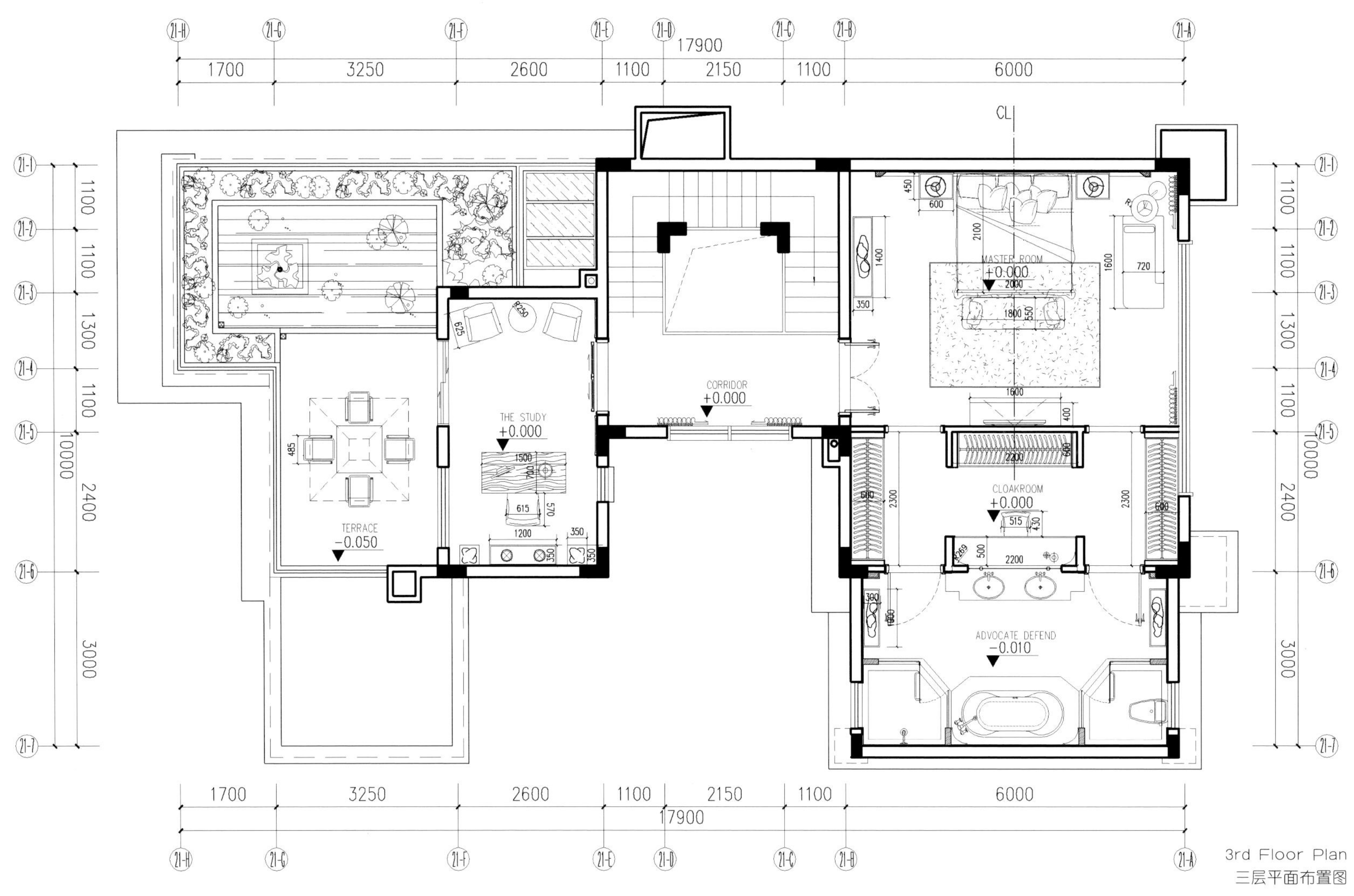

3rd Floor Plan
三层平面布置图

Dragon Pool Villa Neoclassic Show Flat

龙池华府新古典样板房

Designer: Ma Hui,Ge Xulian
Interior Design: Ehe Design Co., Ltd.
Soft Furnishings: Hangzhou Jishang Decorative Design Engineering Co., Ltd.
Location: Suzhou, Jiangsu, China
Area: 430 m^2

设计师：马辉、葛旭莲
室内设计：杭州易和室内设计有限公司
软装设计：杭州极尚装饰设计工程有限公司
项目地点：中国江苏省苏州市
面积：430 m^2

Keywords 关键词

Neoclassicism 新古典主义

Open and Comprehensive 开放包容

Elegant and Luxury 华贵唯美

Furnishings/Materials 软装 / 材料

Gold Beige Marble, Portopo Marble, Coloprhony Onyx, Silver Foil, Ebony Veneers, Wallpapers, Handmade Carpets

金世纪大理石、凡尔赛金大理石、松香玉石材、银箔、黑檀木饰面、墙纸、工艺手工毯

The elegance and luxury of the neoclassicism are fully expressed in this project. Sophisticated designs are used in the limited space to make it more open and comprehensive. No matter it is simple or complicated, overall or locally, all the details are designed carefully. For the accessories, white, gold, yellow and dark red colors are used a lot to create the colorful visual effect. Both the furniture and accessories are elegant and luxury, which have explained the dignity of the master quietly.

Cooperating with the refined European lines, large-sized wallpaper with classic European colors are adopted to define the fashion and taste of European style. In order to simplify the panels, the designers choose the plaster moulding to be the outline.

Stone patterns are used for the floors to soften the artificial feelings by the natural texture and colors, and have showed the luxury and taste of the living room and dining room.

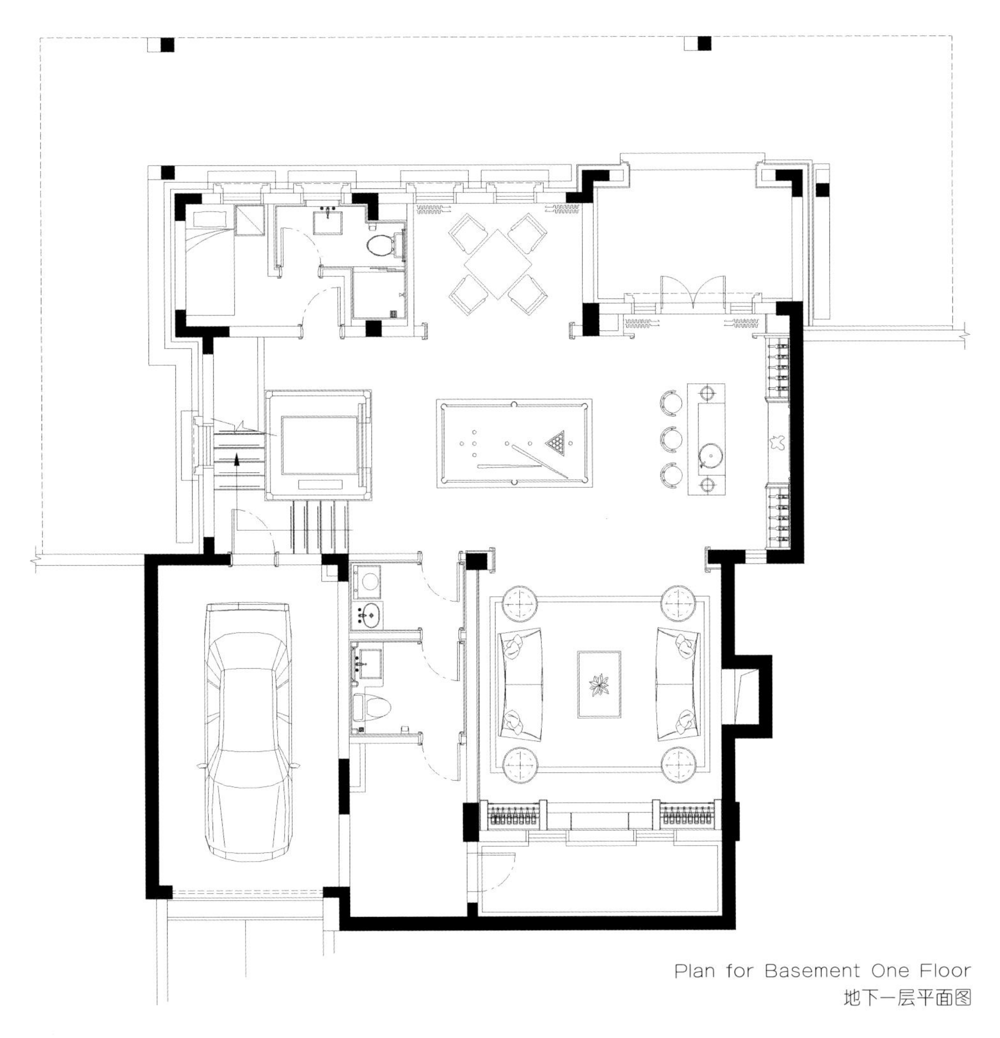
Plan for Basement One Floor
地下一层平面图

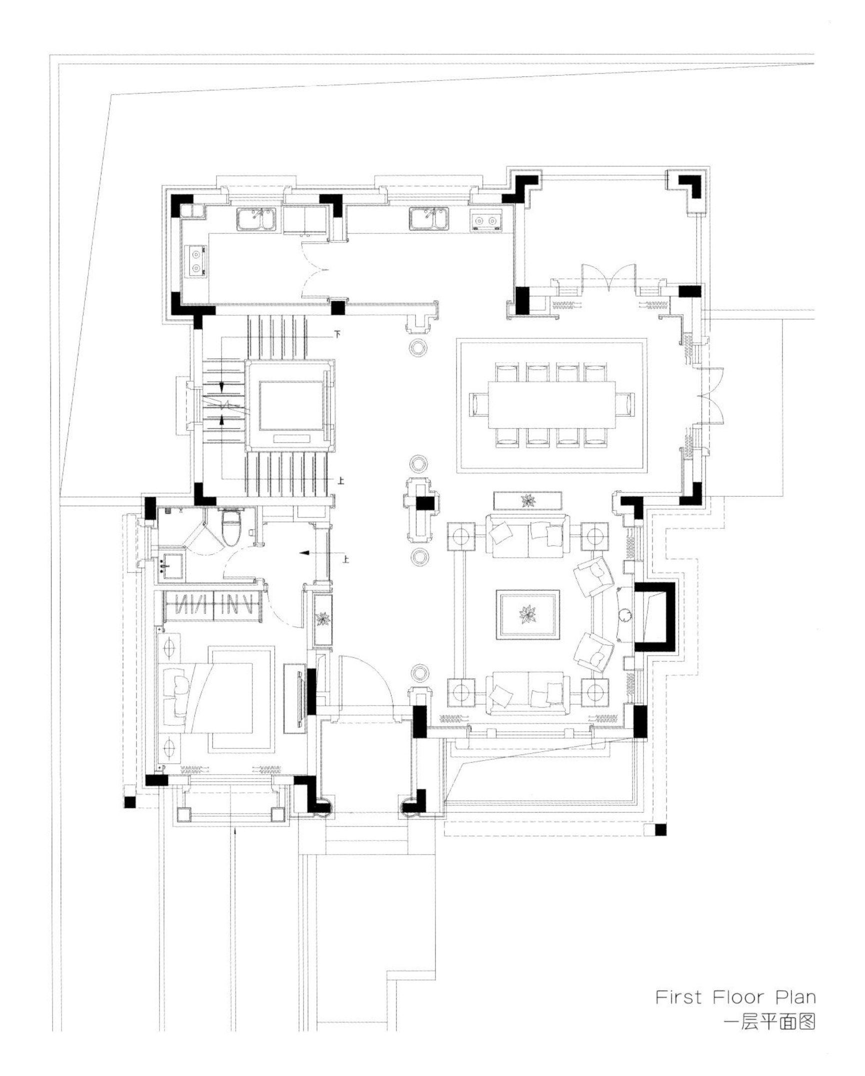
First Floor Plan
一层平面图

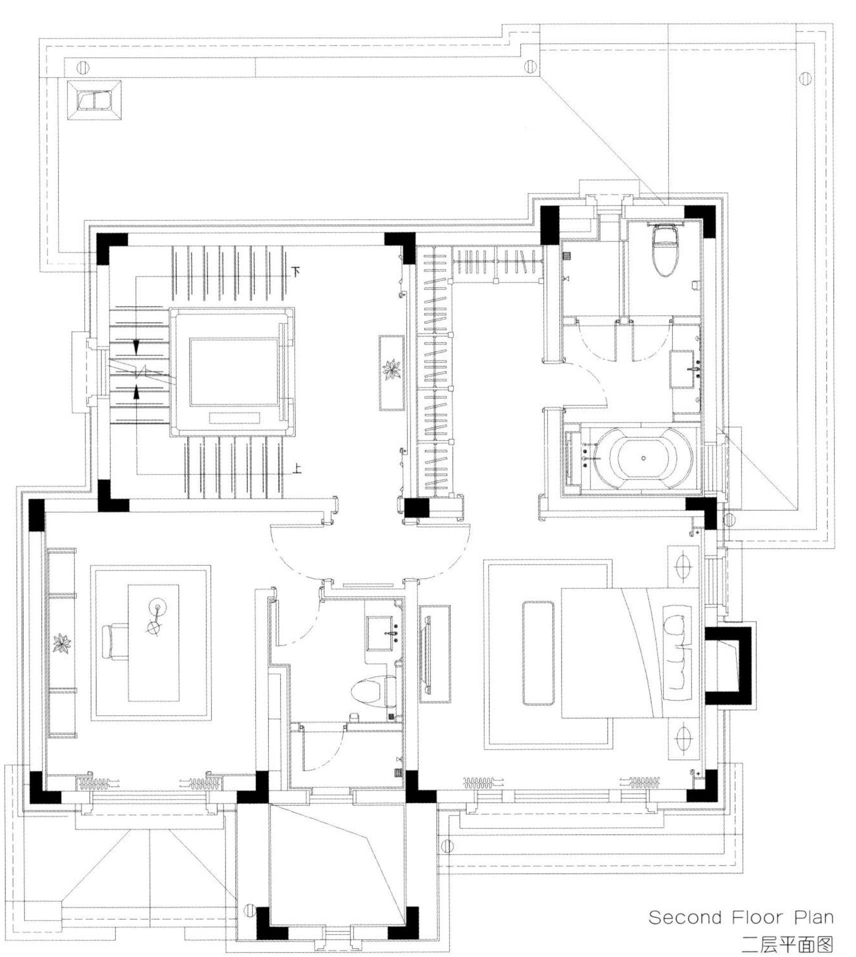
Second Floor Plan
二层平面图

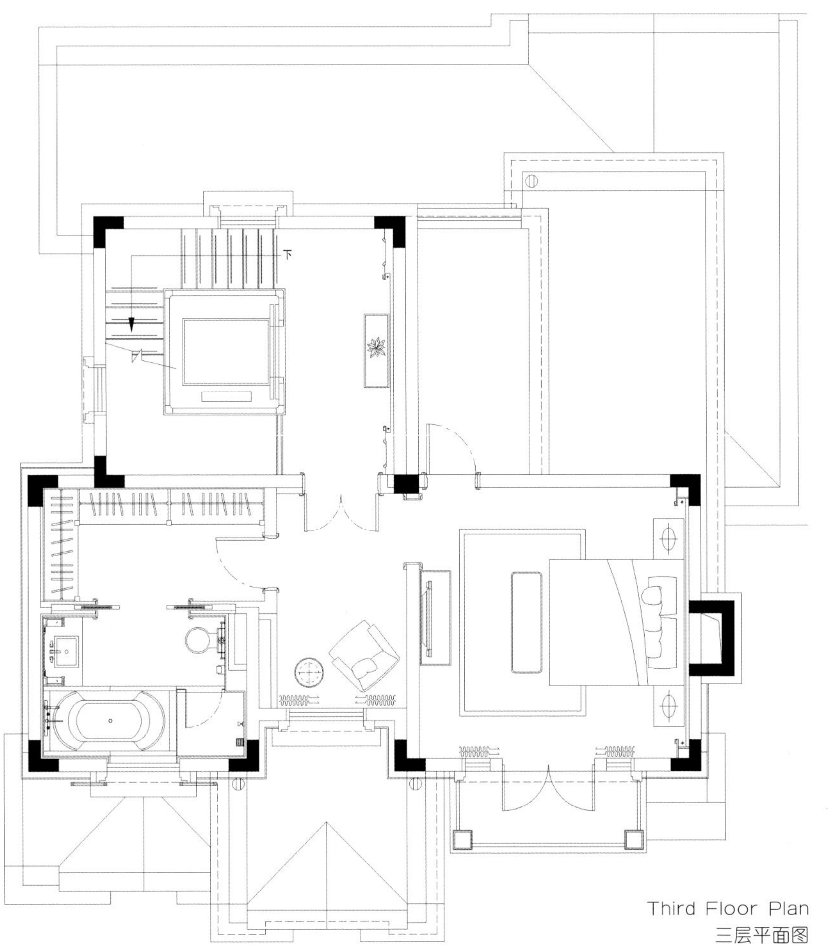
Third Floor Plan
三层平面图

项目的设计风格主要体现出新古典主义的华丽优雅。在有限的空间里，设计师运用一贯擅长的设计手法，让整个空间给人以开放包容的非凡气度，却丝毫不显局促。从简单到繁杂、从整体到局部，都精雕细琢，给人一丝不苟的印象。而在配饰上，使用了白色、金色、黄色、暗红色等色调，着力营造出色彩丰富的视觉效果。无论是家具还是配饰，均以其华贵而唯美的姿态，平和而富有内涵的气韵描绘出居室主人高雅的身份。

墙面大面积地使用了古典欧式色彩的壁纸配合经过提炼的欧式线条，使欧式不再是遥远的过去，而是鲜活时尚的品位象征。不采用复杂的欧式护墙板，设计师使用石膏线勾勒出框线，把护墙板的形式简化到极致。

地面采用石材拼花，用石材天然的纹理和自然的色彩来修饰人工的痕迹，使客厅和餐厅的那种奢华、档次和品位表露无遗。

KELLY HOPPEN
FURNITURE CULTURE
American Style
Mediterranean

The Peach Garden · Shen's Residence, Hangzhou

杭州桃花源沈宅

Designer: Liang Suhang
Interior Design: PMG Interior Design
Location: Hangzhou, Zhejiang, China
Area: 800 m^2

设计师：梁苏杭
室内设计：PMG国际设计机构
项目地点：中国浙江省杭州市
面积：800 m^2

Keywords 关键词

Noble and Elegant 高贵雅致

Clear Outline 轮廓明朗

Neo-classical Decoration 新古典家装

Furnishings/Materials 软装 / 材料

Wallpaper, Stone, Paint, Cooper Products, Iron Products

墙纸、石材、涂料、铜制品、铁艺

With focus on the color scheme, the designer has drawn the outline of the staircase with black lines. The floor is designed with traditional square tiles. No matter the decorations on the table or the bonsai at the corner, every detail shows a kind of vitality.

The walls in the living room are designed to be functional and decorative. To avoid emptiness and monotonousness in the spacious space, the designer has carefully delt with some important nodes to create a wonderful space experience.

In terms of the study room design, it gets rid of the gorgeous decoration, and instead, uses clear outline to highlight the characteristics of the space and create a noble and elegant atmosphere. Simple and concise decorative lines make the space feel comfortable and elegant. When entering the study, people will soon calm down and get the inner peace.

The furnishings in the bedroom directly reflect the lifestyle and the attitude towards life that the designer wants to present and explain to people. When designing the open spaces like the living room, the dinning room, the family room and even the study room, the designer needs to consider the owner's social groups and set the furnishings accordingly. While for the private bedroom, the only concern is the owner's habits and preferences which will help to create a comfortable living space.

The children's room is designed with neoclassical French-style furnishings to tell a fairy tale and create a warm and charming space with the sense of uncertainty.

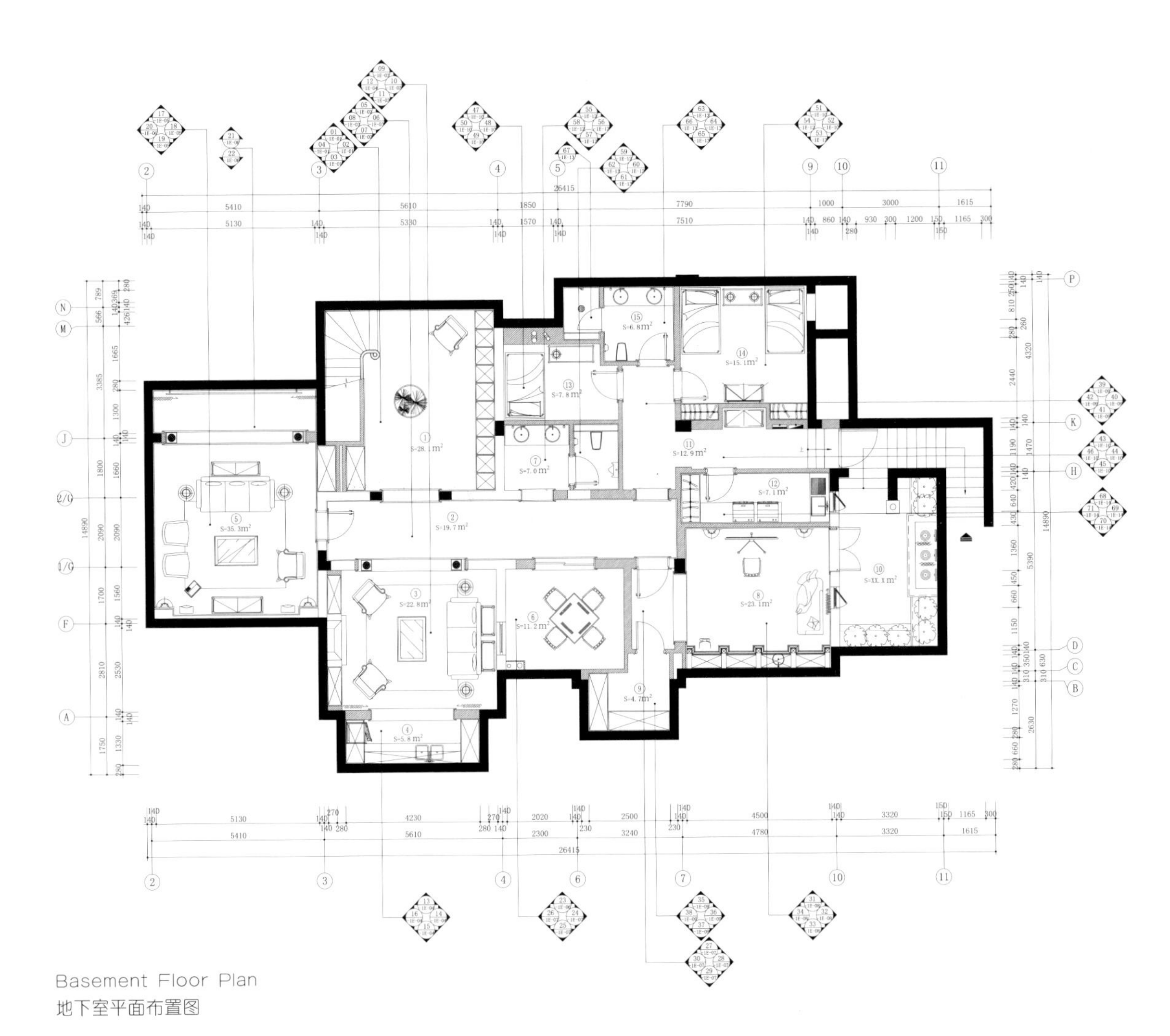

Basement Floor Plan
地下室平面布置图

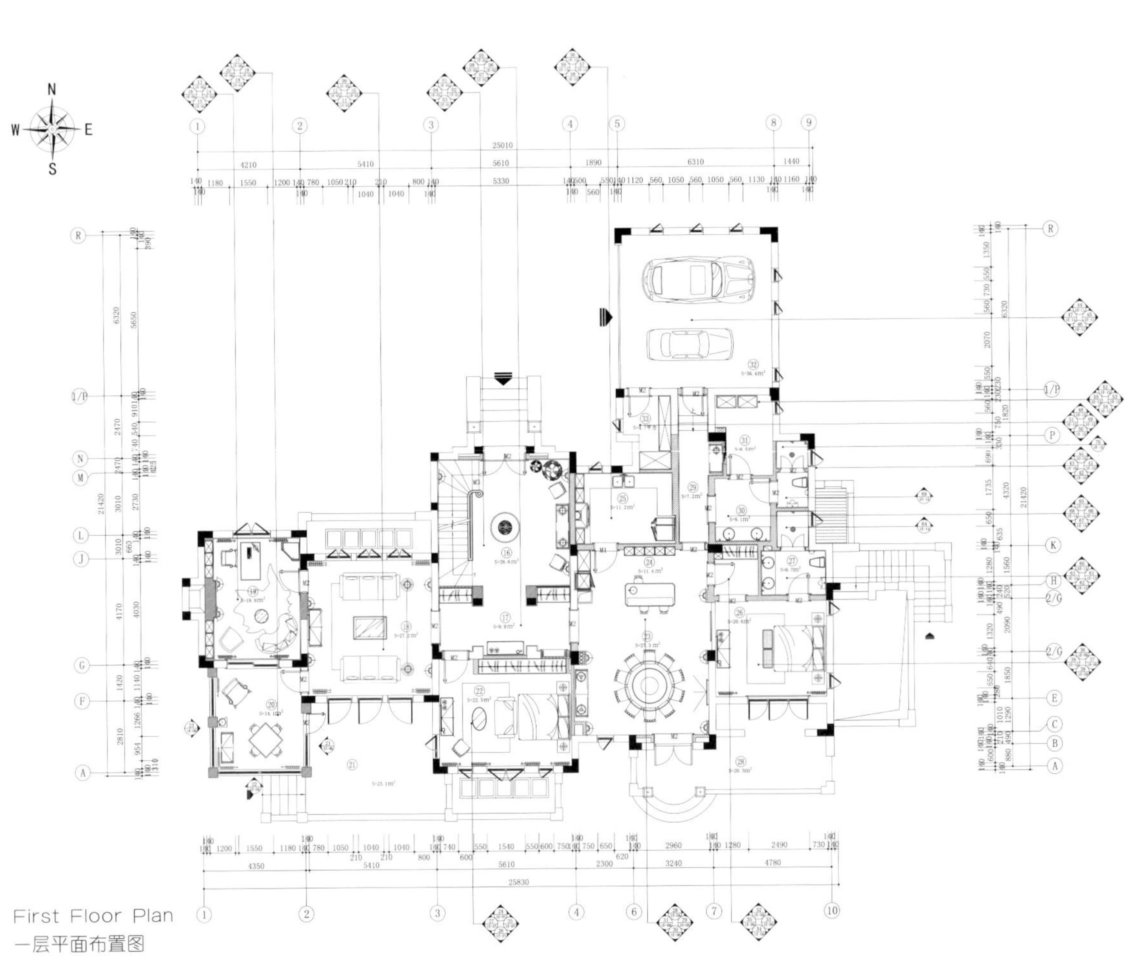

First Floor Plan
一层平面布置图

楼梯间的设计在把握色彩主线的前提下，用黑色线条勾勒出空间的骨架，地板采用传统的格子形式，无论桌上的摆件，还是角落里的盆景，无不展现一种生命力。

客厅的空间墙面处理上以实用性和展示性为主，为了不让充裕的空间显得空旷和单调，在重要的显眼的位置都用心做了一些处理，起到画龙点睛的作用。

在书房的设计上，褪去浓妆艳抹的华丽装饰后，明朗的空间轮廓立刻让这里的特色得到突出，高贵与雅致并存。简洁的装饰线条让整个环境舒适又大气，当人们进去这个空间时，立马沉下心来。

卧室的陈设直接反应出设计师要向人们诠释的生活态度和生活方式。对客厅、餐厅、家庭室乃至书房等开放空间进行设计时，要考虑主人的社交人群这一重要因素来进行陈设设计，但在私密的卧室空间部分则只要最大限度地去考虑主人的特质，基于这一点的基础上来打造一个舒适的生活居所。

儿童房的设计采用法式新古典家装，就像是在诉说一个童话故事，营造出一种既温馨又迷人的未知感。

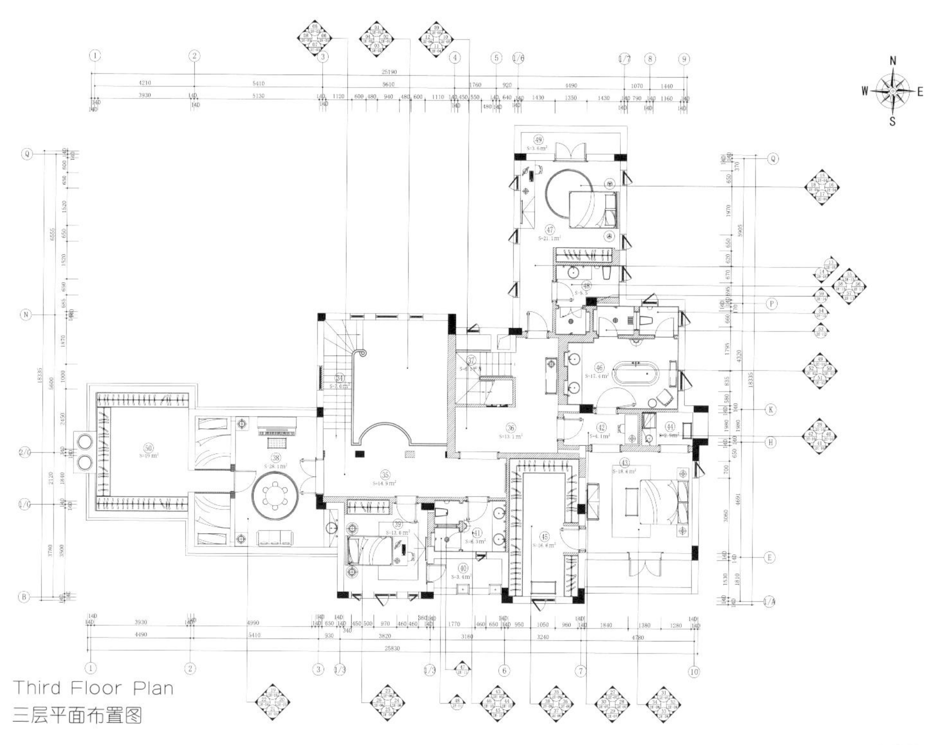

Third Floor Plan
三层平面布置图

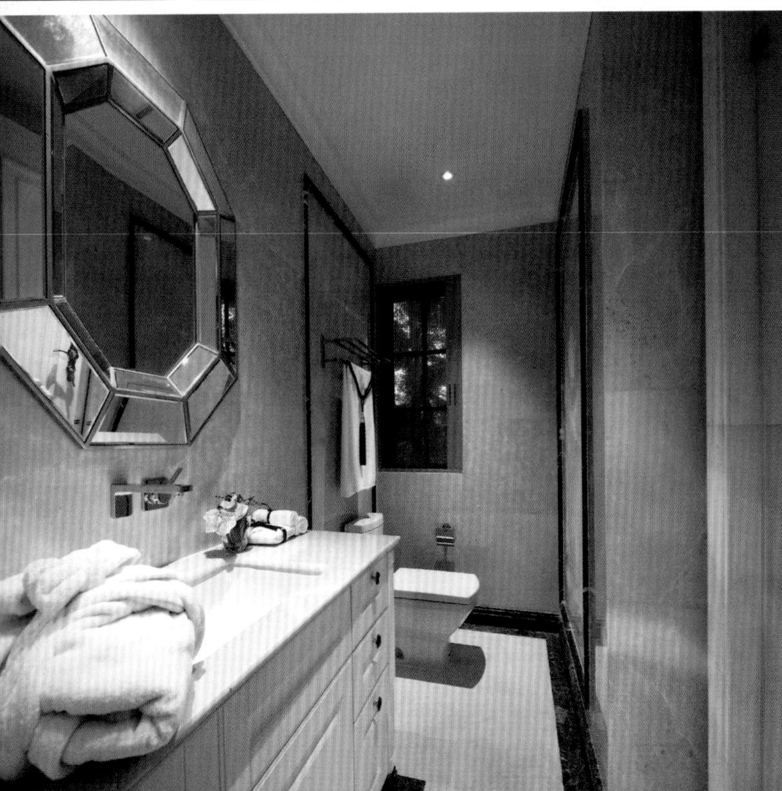

Metropolitan · Guiyang Vintage Style

贵阳 · 中航城复古

Designer: Simon Chong
Interior Design: Simon Chong Design Consultants Ltd.
Location: Guiyang, Guizhou, China
Area: 206 m^2

设计师：郑树芬
室内设计：SCD香港郑树芬设计事务所
项目地点：中国贵州省贵阳市
面积：206 m^2

Keywords 关键词

Retro Style 复古风格

High-ceilinged Design 挑高设计

European Elements 欧式元素

Furnishings/Materials 软装 / 材料

Crystal Chandelier, High-Grade Sofa, Carpet, Crystal Cup, Wine Cabinet, Background Wall, Marble Floor

水晶吊灯、高级沙发、地毯、水晶杯、酒柜、背景墙、大理石地板

By high-ceilinged hall design, the project fully shows the luxurious crystal chandeliers. The classic dark gray high-grade sofa with solid wood frame and the stylish carpet create sense of dignity. Orange Calla Lily quietly on the tea table demonstrates lively vitality. Large glass windows and vertical blinds mix in retro style to create a figure of woman with fluttering and elegant blond.

Soft background wall, elegant big bed and grey brown beddings add certain decorous feeling for the master bedroom on the second floor. The neighboring room for the boy mainly uses camel color. Retro beddings and photo frame on the background wall are symmetrical. While carpet takes vibrant patterns and is embellished with soft furnishings in European elements, adding a little more playful and lively feeling to the room.

Room for the elderly on the first floor also has its own unique design, in which bookcase wallpaper acts as background wall, holding a strong sense of space design, in contrast with the orange beddings to fully demonstrate retro style.

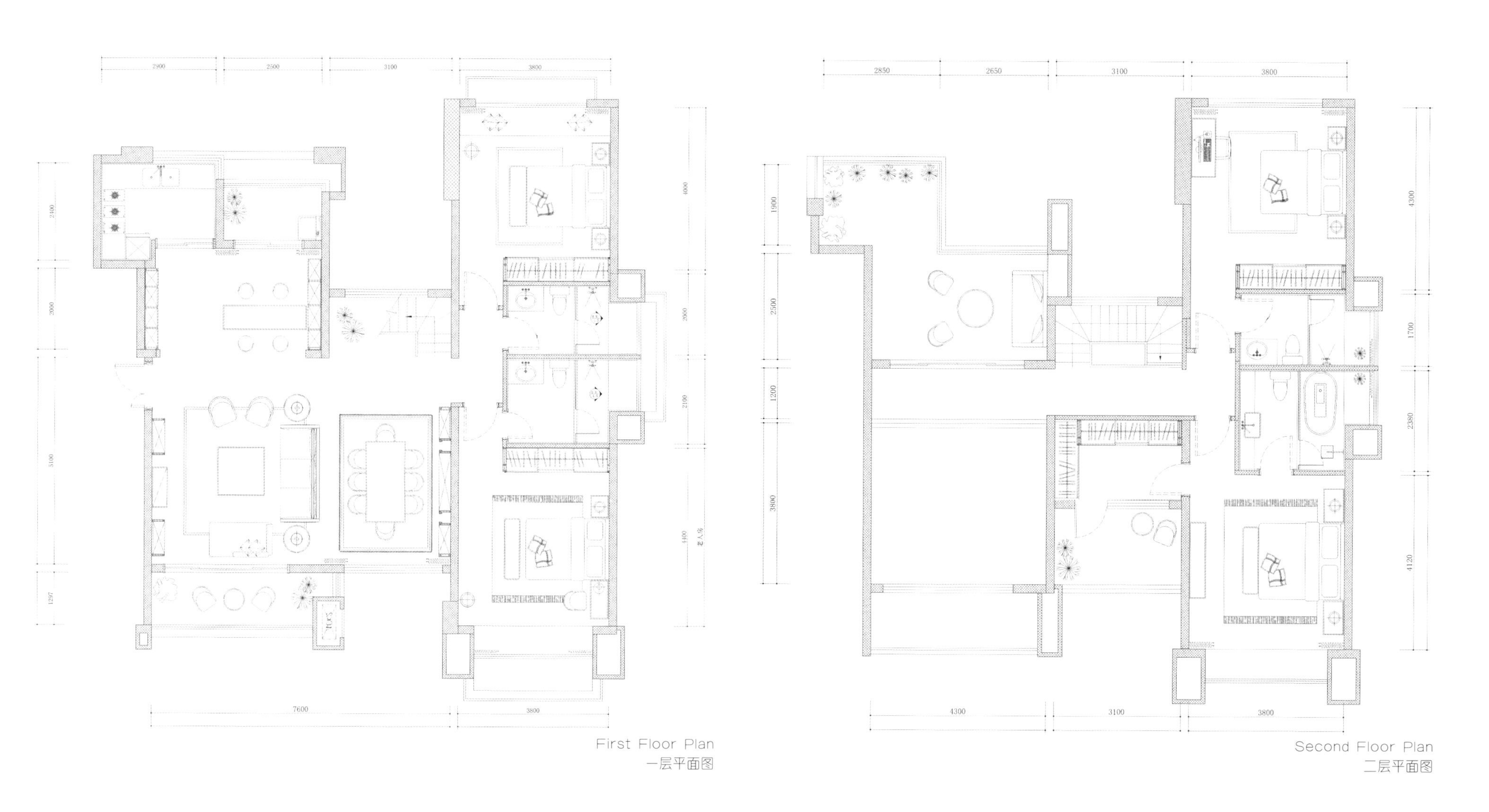

First Floor Plan
一层平面图

Second Floor Plan
二层平面图

项目的大厅采用挑高设计，使得水晶吊灯尽显豪华气派。古典深灰色的实木框架高级沙发与富有时尚感的地毯共同营造出尊贵感。橙色马蹄莲静静地放置在茶几上，展现着蓬勃的生命力。挑空玻璃大窗与垂直的窗帘在复古混搭的意境下塑造出一位金发飘飘的优雅女子。

主卧位于二楼。柔软的背景墙、优雅的大床以及灰褐色的床品为主卧增添了厚重感。紧邻主卧的是男孩儿童房。房间采用驼色为主色基调。复古的床品与背景墙上的相架相互对称。地毯则采用了充满活力的图案及具有欧式元素的软装点缀，使房间增添了几分俏皮和活泼。

一楼的老人房同样有着自己的独特设计。老人房以书柜墙纸作背景墙，空间设计感强，与橘色床品形成对比感，充分展示复古的格调。

SEE
YOU
LATER
FAITH

COFFEE
My box

Take time to pray
3 4 5
COFFEE

My box
Horses

COFFEE
indian

Art Deco Style

Art Deco 风格

Modern Appearance
摩登造型

Clear Layers
层次分明

Mature and Dignified
稳重成熟

Imperial Mansion (House Type A)

北京霞公府 A 户型

Designer: Lian Zhiming
Interior Design: Idee Architecture Interior Design Co., Ltd.
Location: Dongcheng District, Beijing, China
Area: 443 m^2

设计师：连志明
室内设计：北京意地筑作装饰设计有限公司
项目地点：中国北京市东城区
面积：443 m^2

Keywords 关键词

Art Deco Style Art Deco 风格

Low-key Luxury 低调奢华

Female Space 女性空间

Furnishings/Materials 软装 / 材料

Fabric, Flower, Artwork

布艺、花艺、艺术品

The large area of luxurious space and massive use of mirror surface have laid a luxurious foundation of the space. The space is settled as the female living space, so the designer adopts the Art Deco style and renders the space with colors, furniture of international top brand, flower accessories and fabrics.

The main colors of the interior are black, white, beige, and brown, and the secondary and embellishing colors are blue, purple and red. The flexible match and alternate use of various colors make the interior space full of vivacity. In the meantime, plenty of colors are complementary with the furniture, forming an integrated unit.

By using the Belgium vase and the bunchy rose, the extended lines improve the elegance of the project and suggest the grace of the female. The noble but humble clivia, pure white magnolia and so on show the elegance and refine of the female living space and add natural ambiance.

The fabrics are mainly in characteristic skin texture with medium tone to create the ambiance of low-key luxury. The mottled abstract patterns embellish the Turkey velvet and show the taste of the space. The pearlescent purple velvet suggests the female's pursuit of romance. The grayish blue embellishes the bluish violet, creating a quiet harbour for the female host.

项目拥有奢侈的大面积空间，并利用大量的镜面，奠定了空间感的奢侈基础。因为项目的定位为女性的生活空间。设计师结合Art Deco风格，通过色彩、国际一线品牌家具、花艺饰品和布艺渲染空间。

室内主体色以黑色、白色、米色、咖色为主，辅以蓝色、紫色、红色等点缀色。多种色彩的灵活搭配，交替使用，使得室内空间充满活泼的气氛。与此同时，多种色彩与家具进行了互补，形成统一的整体。

通过比利时花器加上束状玫瑰，纵向延伸的线条提升了项目的优雅度，暗示女性的亭亭玉立。高贵又谦虚的君子兰、洁白的木兰花等体现女性生活空间的高雅脱俗，同时增添自然灵气。

布料以选用具有个性肌理配中性的色调为主，营造低调奢华的氛围，斑驳抽象的花纹点缀土耳其绒彰显空间的品位，珠光紫的丝绒暗示女性追求浪漫，灰蓝点缀蓝紫色为女主人打造一份宁静的港湾。

Imperial Mansion (House Type C)

北京霞公府 C 户型

Designer: Lian Zhiming
Interior Design: Idee Architecture Interior Design Co.Ltd
Location: Dongcheng district, Beijing, China

设计师：连志明
室内设计：北京意地筑作装饰设计有限公司
项目地点：中国北京市东城区

Keywords 关键词

Art Deco Style Art Deco 风格

Men's Taste 男士品位

Tough & Ably 硬朗干练

Furnishings/Materials 软装 / 材料

Furniture, Chinese Style Treasure, Painting, Jade Ornament, Flower Decoration, Carpet

家具、中式风格珍品、书画、玉石摆件、花艺、地毯

The interior design of this project is in Art Deco Style, and the soft furnishings design is positioned as a living space for the successful male host who appears in high-end occasions like Beijing Imperial Mansion. It aims to create a gracious, tough and ably style and to lead the super taste and fashion and low-key luxurious style of men in high-end class.

On the selection of furniture, the designer continues to show the Art Deco style elements, and carefully selects the classic works, inheriting the essence of the noble king design style in the whole interior space. The space mainly uses gray and brown colors and furniture with clear and bright line structures to control the overall tone and ambiance of the interior and to highlight the calm and rational temperament of men.

On the details of soft furnishings, the designer boldly infuses heavy Chinese style treasures, paintings, jade ornaments, and flower accessories with rich zen, exquisite and fashionable modern ornaments and so on into the Art Deco style, showing details in grace, tasting zen in exquisiteness, and highlighting the super taste of the host and the artistic ambiance.

The space uses small area of vibrant colors such as red, bright green and orange. For example, the dining room uses carpet, tables and chairs are embellished with bright green to avoid the whole space being dull and depressing and to show the healthy and shining of the host. The whole space is with extreme charm of the male through the foil of the lighting, the perfect dynamic and static combination, the harmony between different style and texture of soft furnishings in this space.

项目的硬装设计为Art Deco风格，软装配饰设计定位为面向出入北京霞公府这种高端场合的成功男士为一家之主的生活起居空间，旨在营造高端男士阶层大气硬朗干练的风格，引领高端男士阶层高品位、高端时尚、低调奢华的风范。

在家具挑选方面，设计师继续运用了Art Deco风格元素，以悉心挑选的经典作品，将尊贵的20世纪王者设计风范之精髓在整个室内空间传承。空间多运用灰色系、咖啡色系及线条结构明确硬朗的家具来控制整个室内色调及氛围，突出男性沉稳、理性的气质。

在软装配饰细节中，设计师大胆运用浓郁的中式风格珍品，书画、玉石摆件及禅意十足的花艺搭配精致时尚的现代装饰摆件等融入Art Deco风格中，在大气中沉淀细节，在细致中品味禅意，彰显男户主高端的品位和艺术气息。

而空间随处小面积地运用红色、鲜绿色、橙色等跳跃的色彩，如餐厅点缀着鲜绿色的地毯及餐椅，使整体不至于沉闷和压抑，体现了男主人阳光、健康的一面。整个空间在灯光的衬托下，动和静完美结合，不同风格、不同材质的软装配饰在空间里相互融合，极具男性魅力。

Luxurious European Style

奢华欧式风格

Aristocratic Atmosphere
贵族气息

Luxurious and Splendid
奢华艳丽

High-end and Graceful
高端优雅

Desperate Love · Mountain View Villa BLD. 48

倾城绝恋 · 虞景山庄 48 幢

Designer: You Weizhuang
Interior Design: Zhuang Design - Shangpin Group
Location: Changshu, Suzhou, Jiangsu, China
Area: 350 m^2

设计师：由伟壮
室内设计：由伟壮设计—尚品组
项目地点：中国江苏省苏州市常熟市
面积：350 m^2

Keywords 关键词

European Style 欧式风格

Gorgeous 富丽堂皇

Lordliness 贵族气派

Furnishings/Materials 软装 / 材料

Gypsum Line, Wallpaper, Emulsion Paint, Soft Cover, Marble, Cut Glass

石膏线、墙纸、乳胶漆、软包、大理石、雕花玻璃

The whole space takes yellow and white as the main color. The European style has no gorgeous decoration and strong colors, presenting a fresh, elegant and generous high-end style, which stands for the fashionable and elegant home design. The TV background wall with texture, the bright and simple curtain and the droplight echoing to the classical carpet together build the European lordliness. The background wall of dining area takes a piece of oil painting. Its flowing lines reflect on the mirror and the TV screen like a dream, making the whole living room more relaxed, pleasant and gracious and presenting an effect that the view changes by moving. The collocation of modern style and European style not only adds the comfort of interior environment, but also presents a typical European world in the living room, making the householder experience the real European atmosphere. Similarly, the mignon wall lamps with paper lampshades matching with the crystal droplight in European style create a distinctive noble feeling.

项目整个空间是以黄色和白色为主。其具有的欧式风格代表了时尚与典雅的家居设计，少了富丽堂皇的装饰和浓烈的色彩，呈现的则是一片清新、典雅和大气并存的高端气派。富有纹理的电视背景墙、明亮朴素的窗帘、和古典的地毯相呼应的吊灯，三者相互配合，共同营造出欧洲贵族气派。餐区背景墙整面采用了一幅油画，流动的线条图案映射在镜面与电视屏幕上，如梦似幻，使整个居室气氛更加轻松、欢快，更有雍容华贵的味道，也显现了移步异景的效果。现代与欧式风格的搭配不仅增加室内环境的舒适感，而且将一个典型的欧洲世界呈现在居室之内，让户主身临其境，感受到实在的欧洲气息。同样，项目的灯饰以欧式水晶吊灯为主，配以小巧玲珑的纸罩壁灯，与众不同的贵族感油然而生。

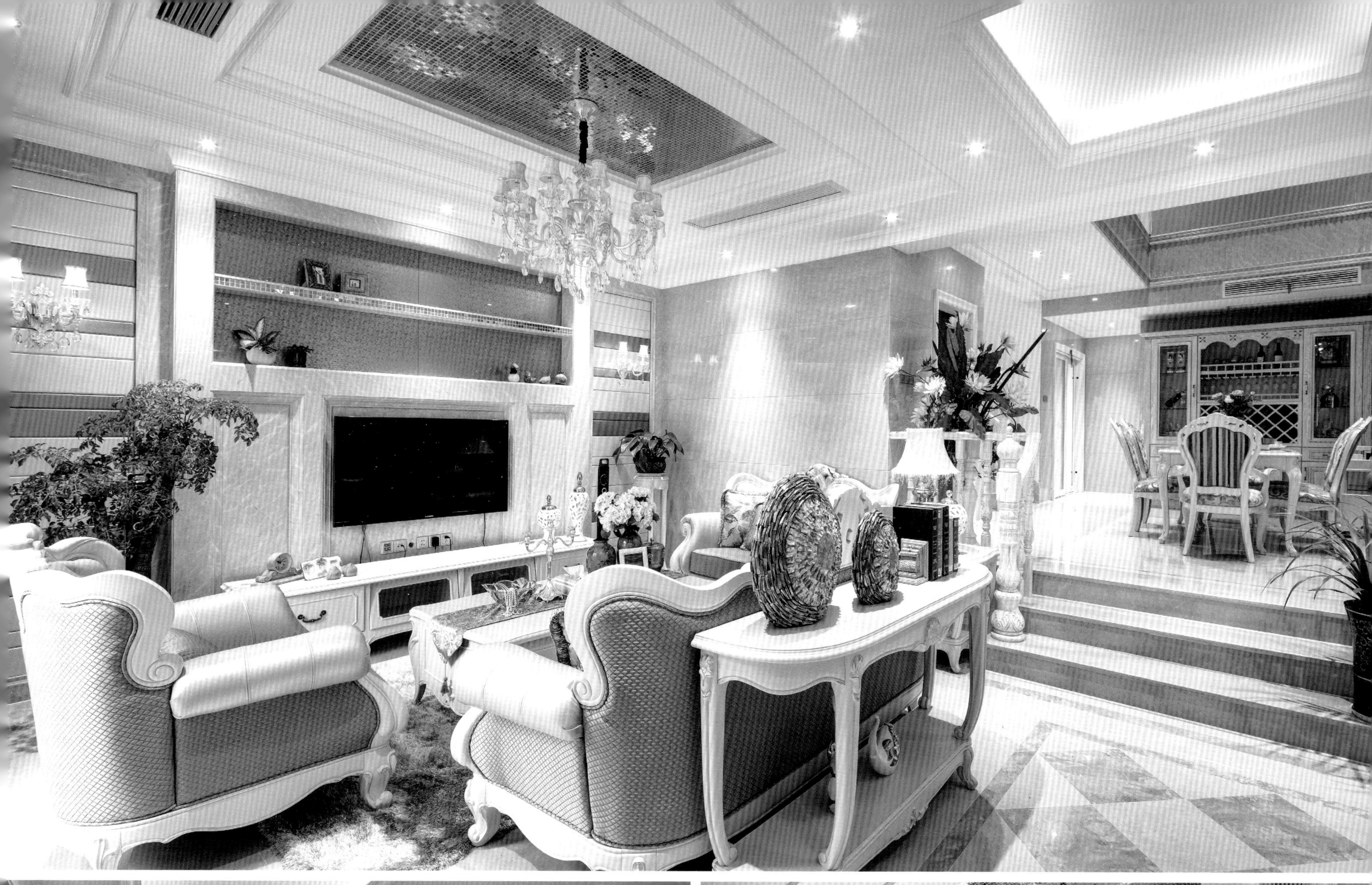

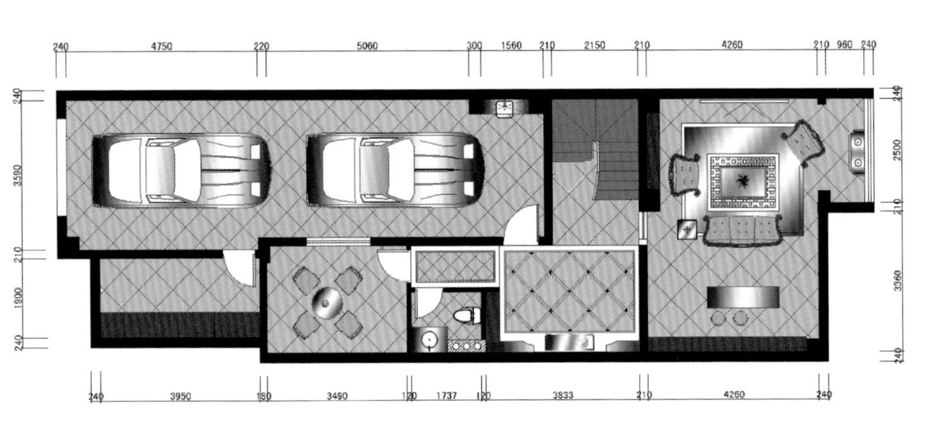

Basement Plan
地下层平面图

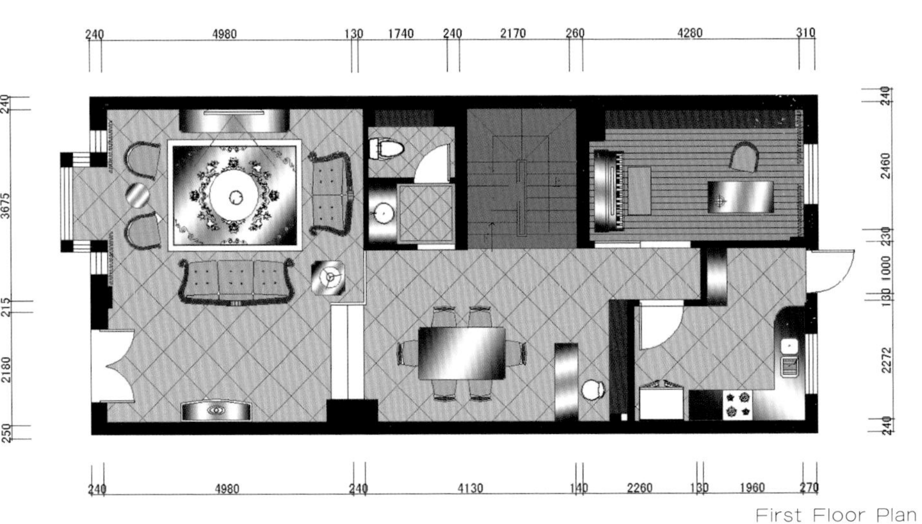

First Floor Plan
一层平面图

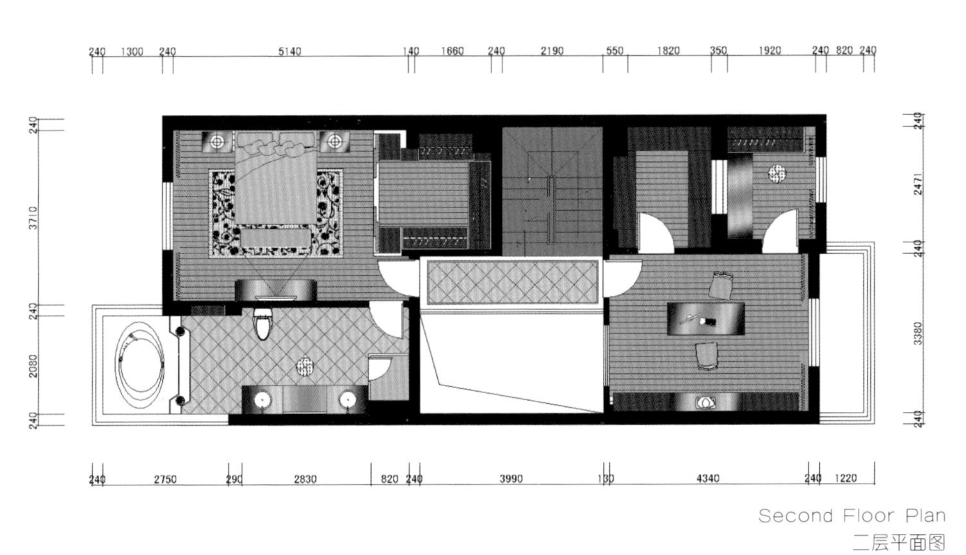

Second Floor Plan
二层平面图

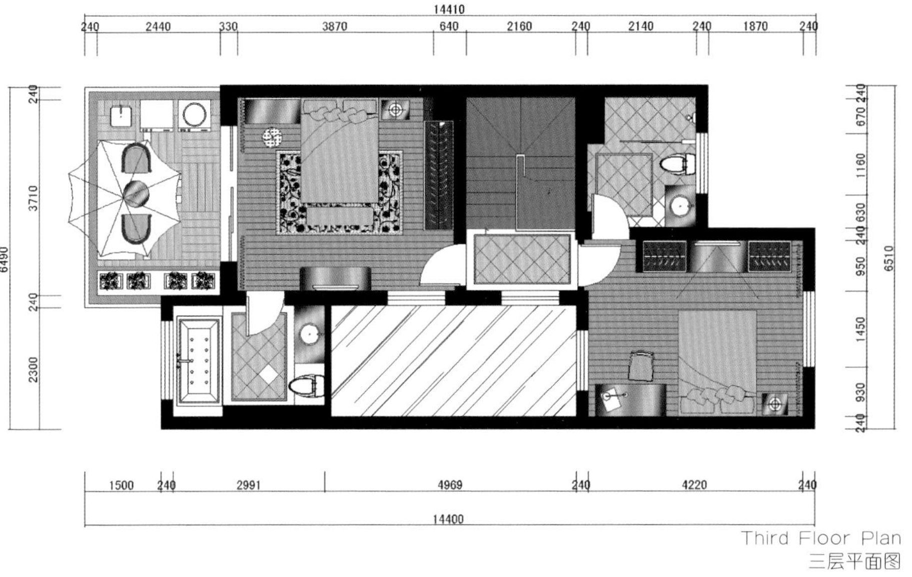

Third Floor Plan
三层平面图

Prince Mansion

华府世家

Designer: Wang Xiaogen
Interior Design: Beijing Genssun International Space Design Co.,Ltd
Location: Dongsheng district, Ordos, Inner Mongolia, China
Area: 560 m^2

设计师：王小根
室内设计：北京根尚国际空间设计有限公司
项目地点：中国内蒙古自治区鄂尔多斯市东胜区
面积：560 m^2

Keywords 关键词

Axial Symmetry 中轴对称式

Dual Hall Area 双厅区域

Photorefractive Effect 光折效果

Furnishings/Materials 软装 / 材料

Stone, Soft Pack, Wallpaper, Mirror, Lobular Rosewood Screen, French Heritage Cabinet

石材、软包、壁纸、镜子、小叶紫檀屏风、法式文物柜

As the first scenery of the interior, the hallway shows the unique ingenuity of the designer. It is in the central area of the whole space, and the inspiration of this functional design evolves from the "ceremony hall" of ancient oriental mansion. The folding squares shaped ceiling is derived from the dome of the Roman Pantheon, and it constantly extends in designs of walls and floors in different space. The overall space layout is in the style of traditional oriental "axial symmetry".

The livingroom and the hallway adopt the concept of "dual hall", which are interconnected, integrated and independent. The couch, tea table and fireplace are vertically arranged in one axis, as well as in the center of the whole "dual hall area". The traditional oriental "axial symmetry" layout displays the solemn breath of oriental culture in this strong luxurious flavored European space; the wall uses materials such as stone, soft package, wallpaper and mirror to form a transparent space with the exterior path of the house, and it perfectly responds to the soft packed walls on both sides of the fireplace in the livingroom; the oriental flavored "lobular rosewood screen" at the entrance responds to the "french heritage cabinet" in the opposite place, showing the integration of two cultures. The facade of the livingroom adopts paintings of oriental cultural theme but in western painting skills. The overall arranging element shows the integration of different cultures by finding a proper joint point between each specialty, and it has infused modern fashionable design elements.

The main bedroom is nearly 15 meters wide with a specially designed Chinese screen that divides it into resting area and study area for effective use of space. Sitting on the soft leather couch, chatting and watching TV become quite comfortable. The study area can be reached just by passing the screen to enjoy the tranquil reading moment in this quiet bedroom. The bathroom in the main bedroom is on the left side of the criss-cross area. The bathroom wall uses bright mirror surface to form a photorefractive effect at the front, both sides and the opposite, and the symmetric mirror design visually extends the space in depth and makes the space more bright and spacious.

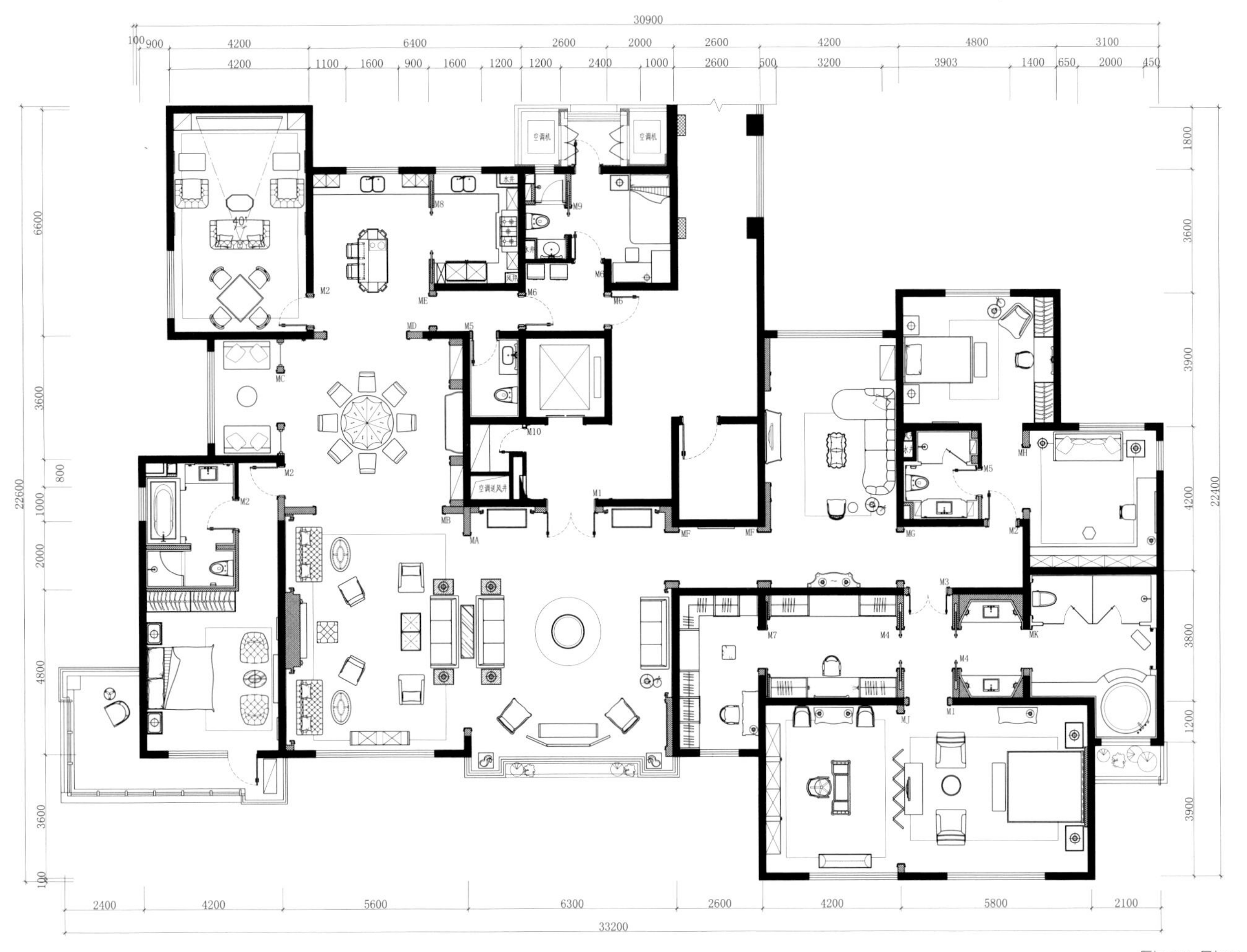

Floor Plan
平面图

作为进入室内的第一道风景，仪式厅是设计师的匠心独运之作。它处于整体空间的中心区域，功能设计的灵感来自东方古老宅邸“仪式厅”的演变。由罗马万神殿穹顶衍生出“方格叠级”的吊顶形式，在不同空间的墙、地面设计上不断延伸。空间整体采用了东方传统的“中轴对称式”布局。

客厅与仪式厅设计“双厅概念”，相互贯通，融为一体，既独立又相连，多人沙发、茶几、壁炉垂直布局于一条轴线，同时也处于整个“双厅区域”的中心位置。这种东方传统的“中轴对称式”布局方法，在这个整体奢华气息浓郁的欧式空间中，展现了东方文明的庄严气息；墙面材质采用了石材、软包、壁纸、镜子等，与家庭室外小走道形成穿透空间，与客厅壁炉两侧软包墙面形成完美对应，虚实互补；入门处一款东方韵味的“小叶紫檀屏风”与正对面的“法式文物柜”遥相呼应，展现了两种文化的交融。客厅的立面运用了西方的绘画技法表现东方文化题材的画品。整体陈设元素均展现了文化之间的交融，在彼此特征中寻求一个恰当的结合点，并融入了现代时尚的设计元素。

项目的主卧室面宽近15 m，特别设计了一款中式屏风将主卧划为“休息区”与“书房区”，有效地利用了空间。坐在柔软的皮质沙发上，可以聊天、看电视。穿过屏风便是书房区域，在安静的卧室可以尽享恬静的阅读时光。主人房“十字形”区域的左侧是“主卧卫生间”。卫生间的正面及两侧利用镜面明亮的“光折效果”作为墙面，对面同样也采取了相同的设计，对称式的镜面设计在视觉上拉伸了整个空间的纵深感，空间变得更加明亮、开阔。

Simple European Style

简约欧式风格

Cozy and Comfortable
惬意舒适

Vivid and Refreshing
清新明快

Vigorous and Vibrant
富有朝气

Medici Palace

天正滨江别墅

Designer: Shen Kaohua
Interior Design: Nanjing SKH Interior Design Studio
Location: Jianye District, Nanjing, Jiangsu, China
Area: 240 m^2

设计师：沈烤华
室内设计：南京SKH室内设计工作室
项目地点：中国江苏省南京市建邺区
面积：240 m^2

Keywords **关键词**

Simple European Neoclassical 简欧新古典

Luxury and Generous 奢华大气

Bright and Concise 明亮简洁

Furnishings/Materials **软装 / 材料**

Ceramic Tile, Ceramic, Lamp Decoration, Solid Wood Cabinet, Block Floor, Custom Furniture, Gold Foil, Silver Foil, Custom Oil Painting

瓷砖、陶瓷、灯饰、实木厨柜、拼花地板、定制家具、金箔、银箔、定制油画

As the luxury villa with five bedrooms, two living rooms, three bathrooms and three balconies, this project is mainly designed in simple European neoclassical style. Designers use all kinds of building materials, such as ceramic tile, ceramic and block floor to build a bright, concise and luxury living atmosphere, which is similar to the royal style.

The design elements of the gold foil honeycomb grille in the living room and the design elements of grille in corridor are echoing. Designers present the space structure completely on material level through the perfect use of materials, and then make creation by "contrast" "coordination" and "unity" on space to increase the luxury and generous elements of the space, such as the faux marble tiles on sofa background wall and the gold foil decoration on the top of the living room. The collocation of wallpaper and the floor, the choice of accessories and the texture of decoration materials, all these dominant elements shape the distinctive temperament of this space.

While in the bedroom, there are the royal queen-sized bed of European style and the delicate beside tables which highlight the extraordinary European style. In addition, the embossment-like flower pattern wallpaper behind the queen-size bed is lively and makes people feel relaxed and happy.

作为拥有五房两厅三卫三阳台的豪华别墅，项目以简欧新古典为主要设计风格。设计师灵活运用瓷砖、陶瓷、拼花地板等各种建材，塑造出明亮简洁而不失奢华的居住氛围，而且颇具皇室气派。

客厅的金箔蜂窝状格栅与过道的格栅造型在设计元素上两相呼应。通过对材料的极致运用，在物质的层面将空间结构一览无余地呈现出来，接着在空间上通过“对比”“协调”“统一”的理念进行创作，比如沙发背景墙的仿大理石砖与客厅顶上的金箔造型，给空间增加了奢华大气的元素。整个空间墙纸与地面的搭配、配饰的选择和装饰材料的质感，这些显性的元素塑造了属于这一空间的独特气质。

卧室采用欧式高挑皇家大床，配以精致的床头柜，尽显欧式的非凡气派。另外，大床背后铺以浮雕般的鲜花图案墙纸，别具一番生机，让人心旷神怡。

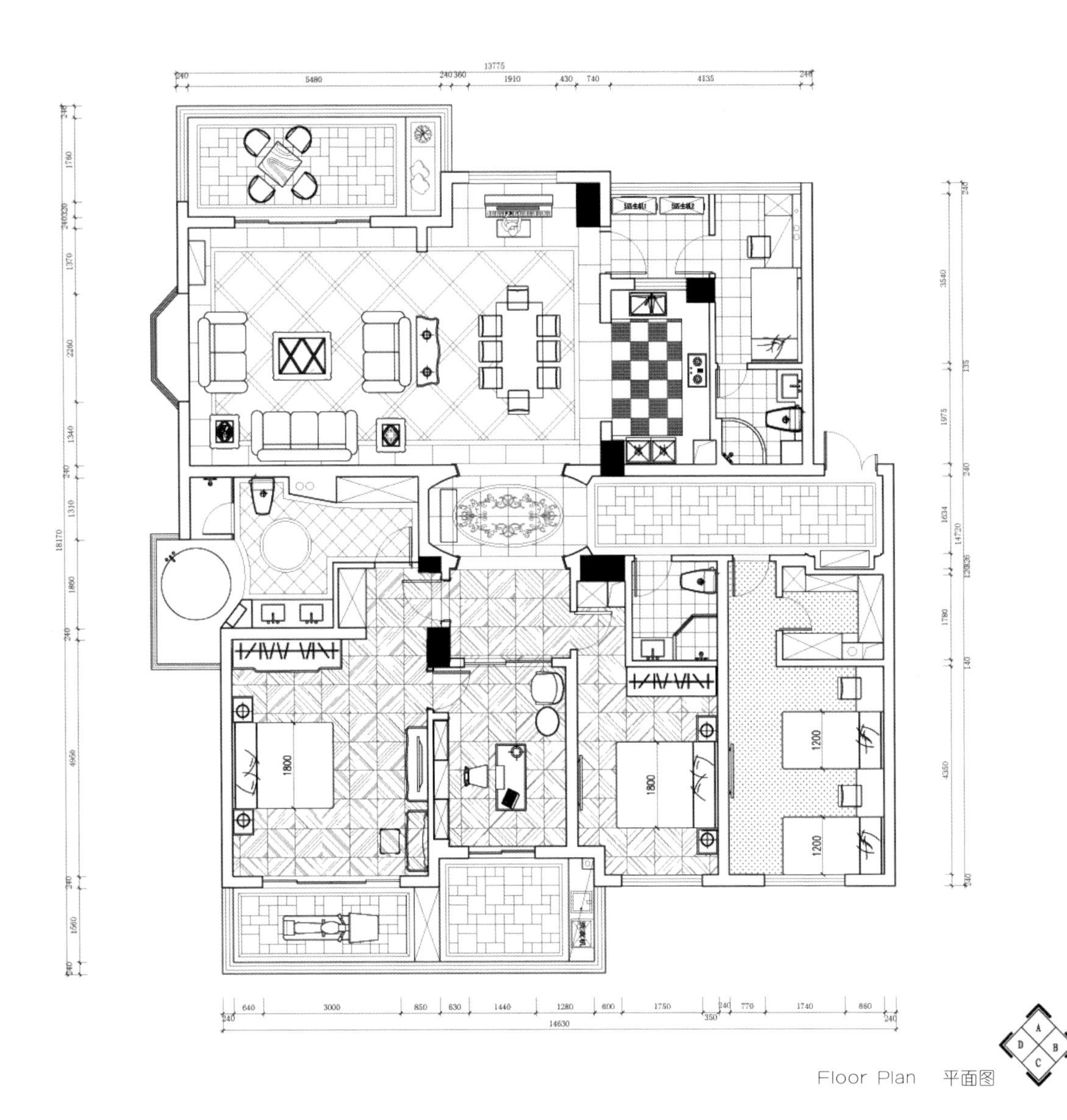

Floor Plan 平面图

GZLakes Show Flat of 7#C-a

广州金地荔湖城 7#C-a 户型样板房

Designer: Lee Jianming
Interior Design: Dongguan Scale Design Co., Ltd.
Location: Zengcheng District, Guangzhou, Guangdong, China
Area: 537 m^2

设计师：李坚明
室内设计：东莞市尺度室内设计有限公司
项目地点：广东省广州市增城区
面积：537 m^2

Keywords 关键词

Simple European Style 简欧风格

Dignified Times 时代凝重感

Comfortable Space 写意空间

Furnishings/Materials 软装 / 材料

Moon Beige, Emperador Dark, Emperor Gold, Silver Foil, Red Sandalwood, Antique Mirrors, Stainless Steel Mirror Surface, Rose Gold Stainless Steel, Wallpaper, Leather

月光米黄、深啡网、帝皇金、香槟银、红檀清香、仿古镜、镜面不锈钢、玫瑰金不锈钢、墙纸、皮革

Adopting simple and concise design style for this project, the designers create a living space that people can feel quiet and comfort, and make full use of the structure of the space to found the exciting and comfortable space. Following the decorative elements of the classic European style, the designers try to achieve the continuity of the space changes and the layering of the volume changes, instead of complicated patterns. They select concise lines, as well as bright and fresh colors, to retain the elegance and luxury of classic European style and also meet the requirements of leisure and comfort for the modern lives. The attitude for the life of the master, such as pursuing for quality and elegant life, regarding the life as an art, can be fully seen from the deluxe design. Lots of decorative materials are used for the unit, for example mirrors, crystal, silver foil wallpaper, stone, wood, tiles and so on, but they are in a unified style and still reveal the concise atmosphere. Besides, large-scale white wooden veneers are set to express this dignified times, as well as the feelings to pursue for a better life.

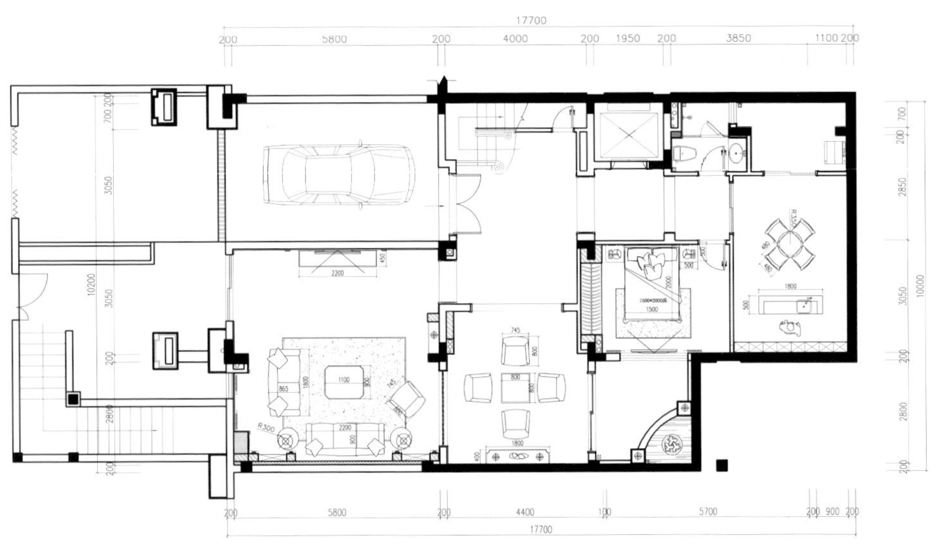

Basement Floor Plan
负一层平面布置图

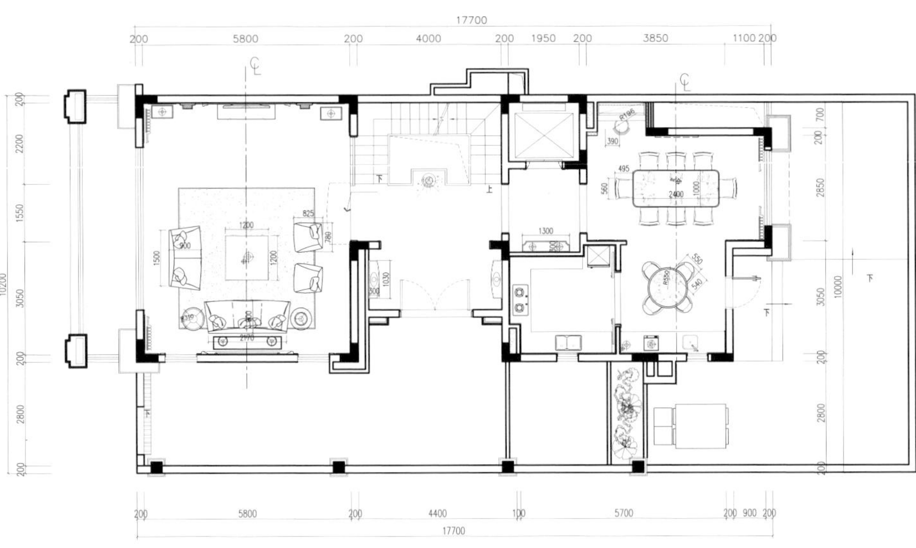

1st Floor Plan
一层平面布置图

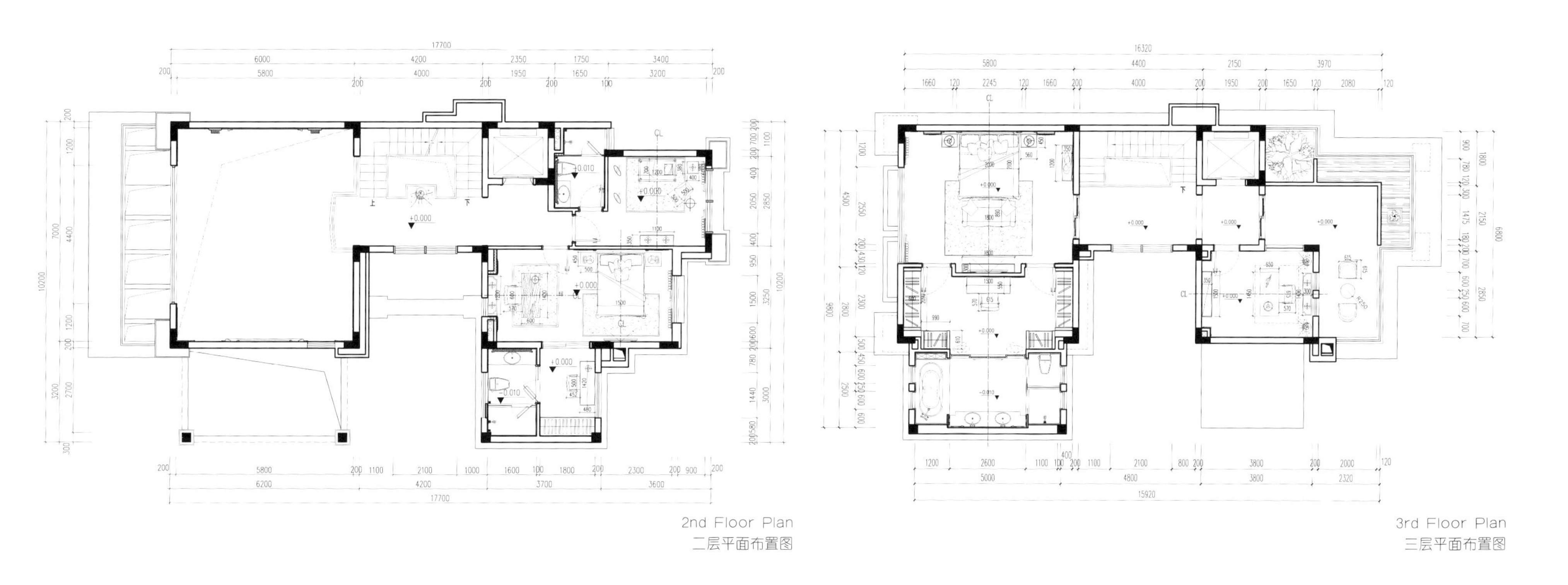

2nd Floor Plan
二层平面布置图

3rd Floor Plan
三层平面布置图

项目采用深受青睐的简约、质朴的设计风格，让人在简单、自然的生活空间中身心舒畅，感受到宁静和安逸。设计师借助室内空间的结构，缔造出一个令人心驰神往的写意空间。项目继承了传统欧式风格的装饰特点，吸取了其风格的“形神”特征，在设计上追求空间变化的连续性和形体变化的层次感。在造型设计上以简约的线条代替复杂的花纹，采用更为明快清新的颜色，既保留了古典欧式的典雅与豪华，又更适应现代生活的休闲与舒适。豪华的设计表现能够完整地体现出居住人追求高品质、典雅生活，视生活为艺术的人生态度。项目虽然使用了诸多装饰材料，如镜子、水晶、壁纸银箔、石材、木材、瓷砖等，但是在整体把控上严谨统一，并未造成繁琐杂乱。同时又以大面积的白色木饰面装饰出时代的凝重感，使人们有了一种对生活迫切追求的感受。

ELLIS ISLAND

Schnauzer
65

Metropolitan · Guiyang Living at Provence

贵阳中航城 · 住在普罗旺斯里

Designer: Simon Chong, Amy Du
Design Company: Simon Chong Design Consultants Limited
Location: Guiyang, Guizhou, China
Area: 103 m^2

设计师：郑树芬、杜恒
室内设计：SCD香港郑树芬设计事务所
项目地点：中国贵州省贵阳市
面积：103 m^2

Keywords 关键词

Low-key and Steady 低调内敛

French Romance 法式浪漫

Pink and Soft Grace 粉色柔美

Furnishings/Materials 软装 / 材料

Upholstered, Shallow Wooden Veneer, Chair, Leather and Rivet Furniture, Bedding, Hanging Paintings, Carpet, Crystal Chandeliers, Pattern Wallpaper

软包、浅木色墙面、单椅、皮革加铆钉家具、床品、挂画、地毯、水晶灯、花纹壁纸

A delicate and romantic living atmosphere in Provence style is decorated by the designer in this 103 m^2 space, which includes 3 rooms and one of them is study room. Soft living room and master bedroom, innocent and romantic child's bedroom, as well as sagacious study room are cooperating together to create a low-key but also romantic atmosphere which is enhanced layer upon layer, making it possible for the master to fully enjoy the flavor of French country.

By adopting veneer in walnut color for the whole living and dining rooms to express unique style, the choose of wooden fabric furniture and hanging abstract paintings have brought the space into the flavor of French country, together with the decoration of square art glass and crystal lamps, a spacious and luxurious sense can be improved. With the help of placed flowers and plants, enormous vitality can be felt in both living and dining rooms. In a word, it makes it possible that master can enjoy the charm and romance of Provence even though at home.

Strong French flavor can not only be enjoyed in the living room, but also in the master bedroom, such as the soft veneer in light wooden color, pink bedside and chair, as well as white leather and rivet furniture. Moreover, hanging paintings, refined carpet and crystal

chandeliers have also represented the elegant and luxury flavor of French country to the master.

While different from the master bedroom which is in graceful and soft style, children's bedroom selects green pattern for both wallpaper and bedding to reveal the low-key and steady quality, and to involve the characteristics of pureness and fun through alphabetic furniture and interesting decorations.

Various kinds of sculptures are placed in the study room, which are not strange at all but expressing master's pursuit of the artworks. While the using of wooden furniture and bookcase, gray pattern wallpaper and ancient phonograph have brought master back to the French country. The ancient carpet along with a lot of collection of books has created a space for independent thinking and for intoxication of French flavor.

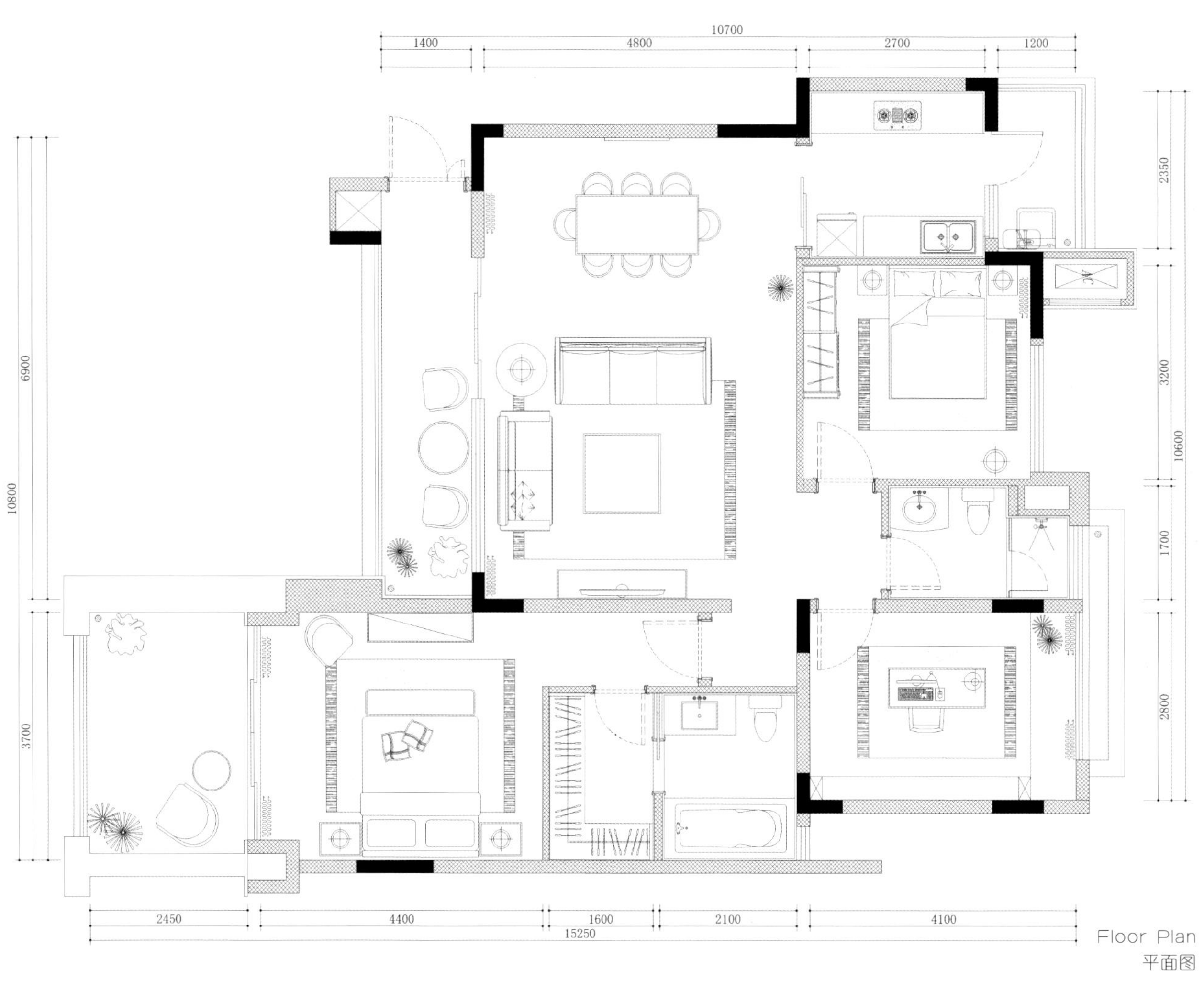

Floor Plan
平面图

设计师在103 m^2的空间中装扮出一个精致而浪漫的普罗旺斯生活意境。项目共有三个房间，其中一个为书房。客厅和主卧的柔美、儿童房的天真浪漫、书房的睿智共同营造出层层叠进的低调、内敛、浪漫气息，让户主充分体验法式乡村的风情。

桃木色饰墙面环绕了整个客厅和餐厅，突出了不一样的格调。而木质的布艺家具和悬挂的抽象画让整个氛围沉浸入了法式乡村的浓浓气息，再加上方格艺术玻璃和水晶灯具的照耀，整个空间感提升，同时具备奢华格调。而在花儿和绿化的点缀下，生命力也洋溢着整个客厅和饭厅。总而言之，在家里也能感受到浪漫的普罗旺斯魅力。

浓重的法式风情不止体现在客厅里。在主卧里，一丝柔美的浅木色墙面、粉红色的床头、单椅和白色皮革加铆钉家具体现着浓浓的法式浪漫风情。挂画、雅致的地毯、水晶灯让精致而奢华的法式乡村风情呈现在户主的眼前。

儿童房的风格却不同于主卧那股优雅柔美的女人风情。绿色花纹壁纸和床品都令空间表现出一份思考内敛的气质。而字母家具和趣味的装饰品让整个空间不仅尽显低调，而且饱含一份纯真童趣。

书房里各种雕塑有条不紊地摆在各个位置，非但不显突兀，反而体现出了户主对艺术品的追求。而木色家具和背景柜、灰色花纹的壁纸、复古留声机又让户主回到法式乡村。复古地毯和满满的收藏书籍让整个书房不仅呈现独立的思考空间，更令人沉醉在法式风情中。

A Treasury of World's Classics Marla Stuart Stafan Zweig
CHARLES DARWIN THE DESCENT OF MAN
Heshi
European classical ancient empire
European-style selection
Empire classic books
THE COUNT OF MONT CRISTO

Haide Park Residence

海德公园住宅

Designer: Huang Shuheng
Interior Design: Sherwood Design
Location: Xinzhuang District, Xinbei, Taiwan, China
Area: 238 m^2

设计师：黄书恒
室内设计：玄武设计
项目地点：中国台湾新北市新庄区
面积：238 m^2

Keywords 关键词

Classical Glamour 古典气派

Steady & Clean 沉稳洁净

Orderly Totem 图腾次序感

Furnishings/Materials 软装 / 材料

White Piano Paint, Glossy Ebony Veneer, Low Formaldehyde Waterborne Wood Coating, Imported Futures Wallpaper

白色钢琴烤漆、亮面黑檀木皮、低甲醛水性木器漆、进口期货壁纸

Stepping into the hallway, one is greeted by the golden brown wall with carved patterns. The round complicated flower ornaments together with the vertical lines manifest the exquisiteness of the classical royal family heraldry. The big glossy mirror and the lamps that can reflect black and golden streamlines are the preliminary echo to historical and modern sense. While walking and moving slowly, one can feel that the space of the living room is quite different from the solemn and rigorous hallway. It maintains the classical sense in details, such as the golden totems at both sides of the TV cabinet. The various light colors in the living room such as light blue, light green and bright silver are soft and pleasant. The grass green chair with rich natural flavor, bright yellow curtain and decorating materials in Victorian style express the vitality of the space and show the family status and unique taste of the female host. While the dinning space, a happy gathering place for relatives and friends, should be in a steady and clean atmosphere. The designer selects cross grained light brown siding as background and places a big painting on it, which is like a rippled round line board that hangs a crystal chandelier. The light color and the platter porcelain below jointly create an elegant taste.

The study room keeps the light gold as it always has, and it uses the simple lined bookcase as theme to extend the view to the white desk and chair. The crystal lamp on the desk echoes far with the dinning hall. For the bedroom space, the designer uses hidden skills to create massive storage space at the host's request. The main and secondary bedroom use the consistent white background, and collocate it with blue and green lines in different proportion, bright decorative patterns scattered at the curtain, throw pillow and carpet, slightly separate it with golden brown and show the order with repetitive classical totems, showing the family relationship model of seeking commonness among differences. And this is also the core meaning of "modern life".

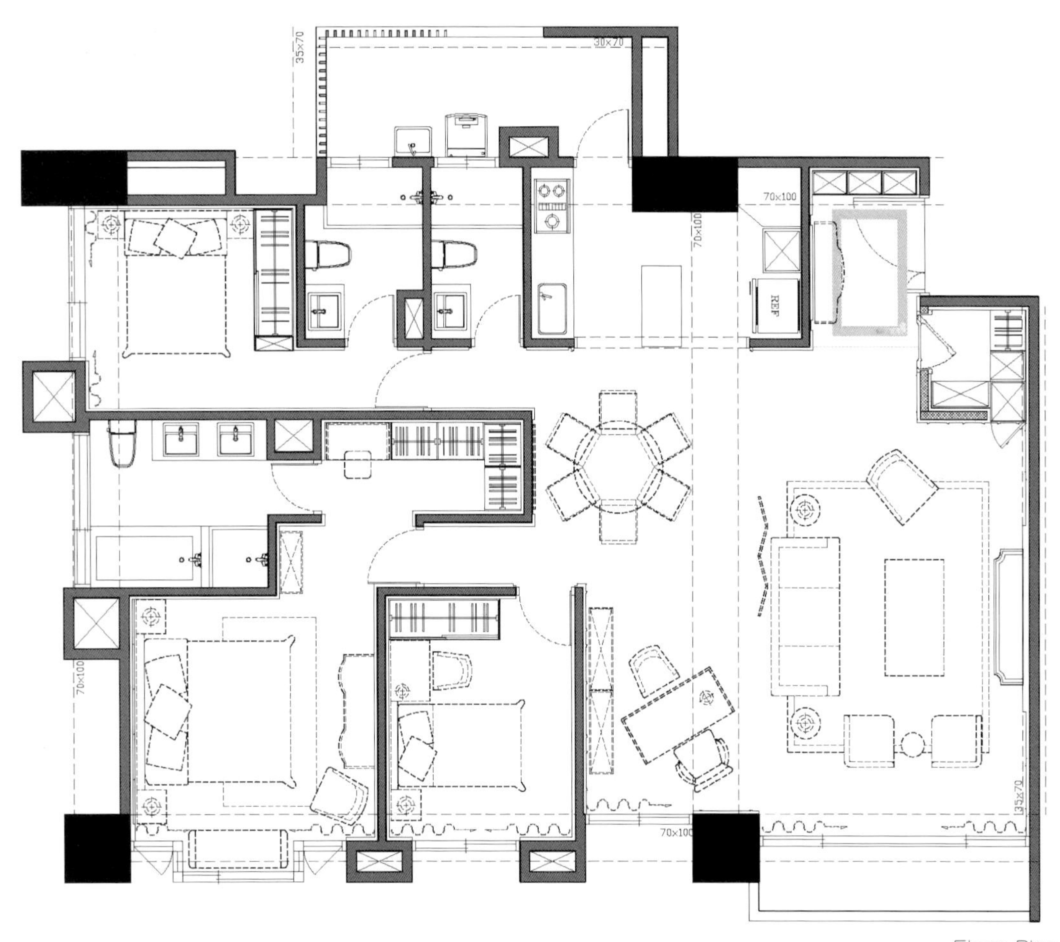

Floor Plan
平面图

MODERN GLAMOUR

Mixed Chinese-western Style

中西混搭风格

Sino-west Mix

中西合璧

Simple and Elegant

古朴典雅

Beautiful and Sophisticated

美观精致

Changsha CSC Meixi Lake Era Villa Sample Room

长沙中建梅溪湖一号别墅样板间

Designer: Liu Weijun
Interior Design: PINKI Creative Group & Liu and Associates (IARI) Interior Design Co., Ltd.
Location: Yuelu District, Changsha, Hunan, China
Area: 350 m^2

设计师：刘卫军
室内设计：PINKI品伊创意集团&美国IARI刘卫军设计师事务所
项目地点：中国湖南省长沙市岳麓区
面积：350 m^2

Keywords 关键词

Magnificent 华美富丽

Graceful and Romantic 柔美浪漫

Mysterious Atmosphere 神秘气息

Furnishings/Materials 软装 / 材料

Rosy Clouds Shadow Wood Veneer, Italian Wood Grain Marble, Silver Dragon Marble, Rose Gold Stainless Steel

云霞影木饰面、意大利木纹大理石、银白龙大理石、玫瑰金不锈钢

This project creates a new Asian style by integrating the western art culture into the oriental traditional way of life, which is a collision between the east and the west and makes the theme way of life present a neoteric visual effect. Warm private room lays emphasis on enjoyment. This craft offers people a new visual shock and new thinking shock (theme presentation).

The layout takes space as an integrated system which is human-centered and includes the universe. Every integrated system in this environment is an important element on interrelation, interaction, interdependency, mutual opposition and mutual transformation. Function layout is grasping the relationship between each subsystem and optimizing the structure to seek the best combination.

This project takes concise and clear lines and elegant and decent decoration to present the magnificent atmosphere of the space. This kind of design expresses a kind of relaxed and comfortable style, which makes the house a place to release pressure and relieve fatigue, and offers people an elegant, quiet and solemn sensual pleasure. The concise ceiling design and natural wood grain floor seize the essence of simplicity, symmetry and elegance in American style, while the use of white marble pillars in the space expresses a kind of spirit which is more rational, balanced, pursuing freedom and advocating innovation.

The magnificent curtain matched with the crystal droplight brings a graceful and romantic atmosphere for the space which takes hard lines as the main element; while the ceiling figures used in the

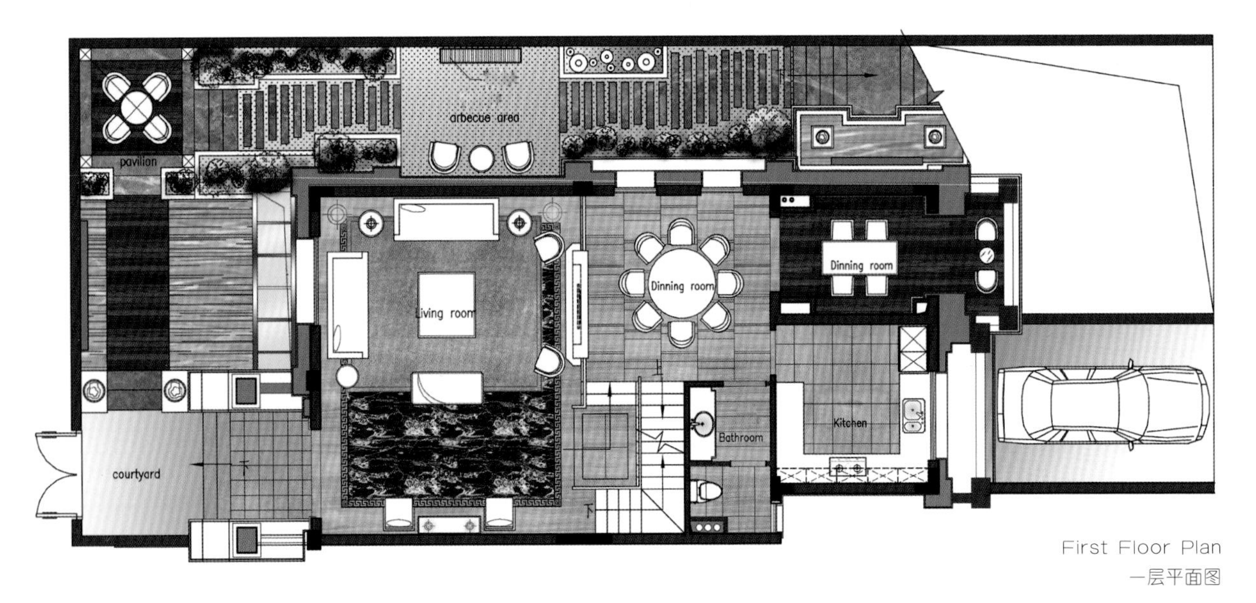

First Floor Plan
一层平面图

master bedroom and the multi-function room like the flowing clouds in the sky, which add some mysterious atmosphere in rational balance.

Sitting on the large primitive and steady sofa and chairs reading or drinking mellow red wine all can make you feel comfortable and relaxed at home and free yourself from the five-star hotel or the office. The light blue and pink decorations add a bright color on the steady puce tone, while the tinsel candlestick and the side cabinet with carved patterns in the living room and the dining room present a unique charm.

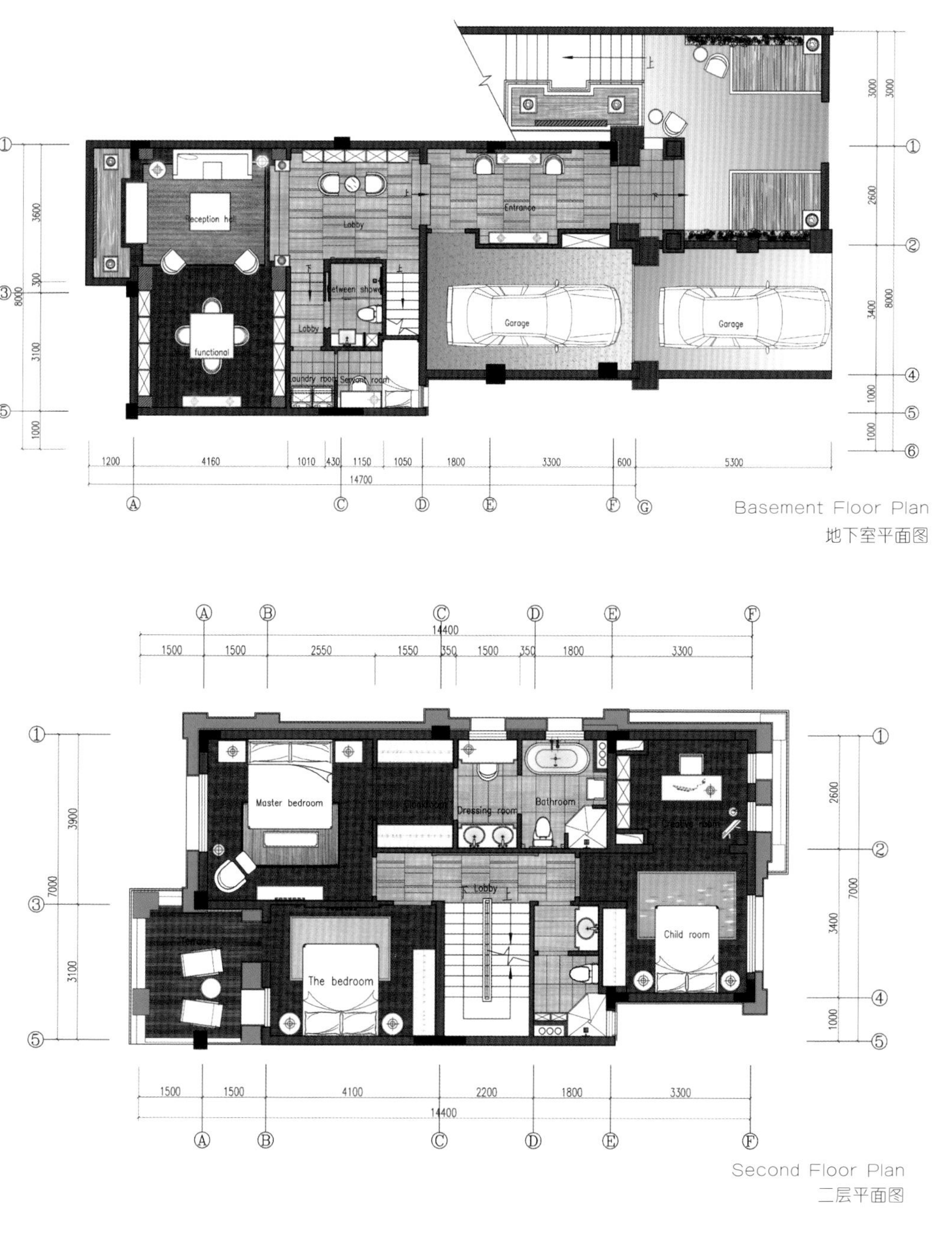

Basement Floor Plan
地下室平面图

Second Floor Plan
二层平面图

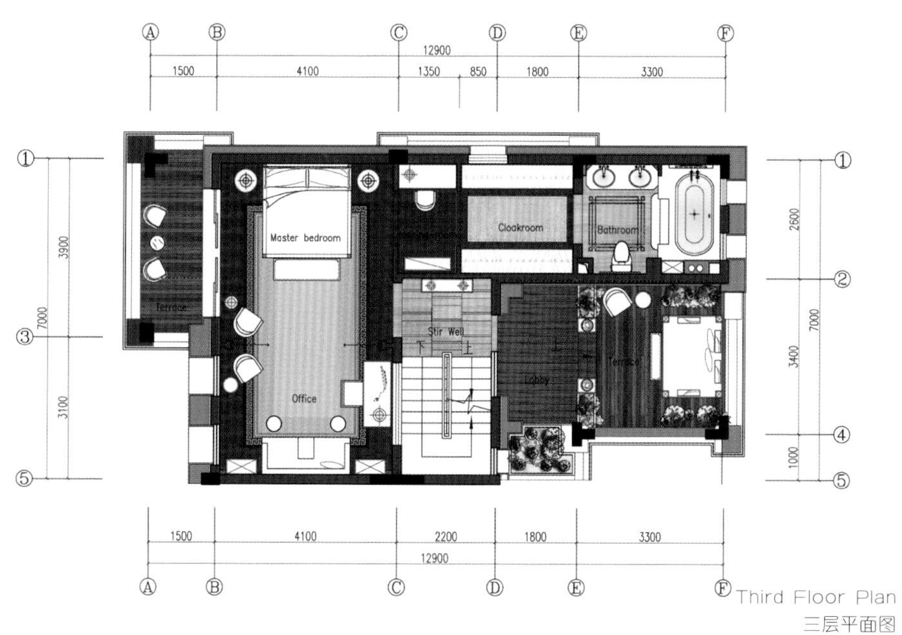

Third Floor Plan
三层平面图

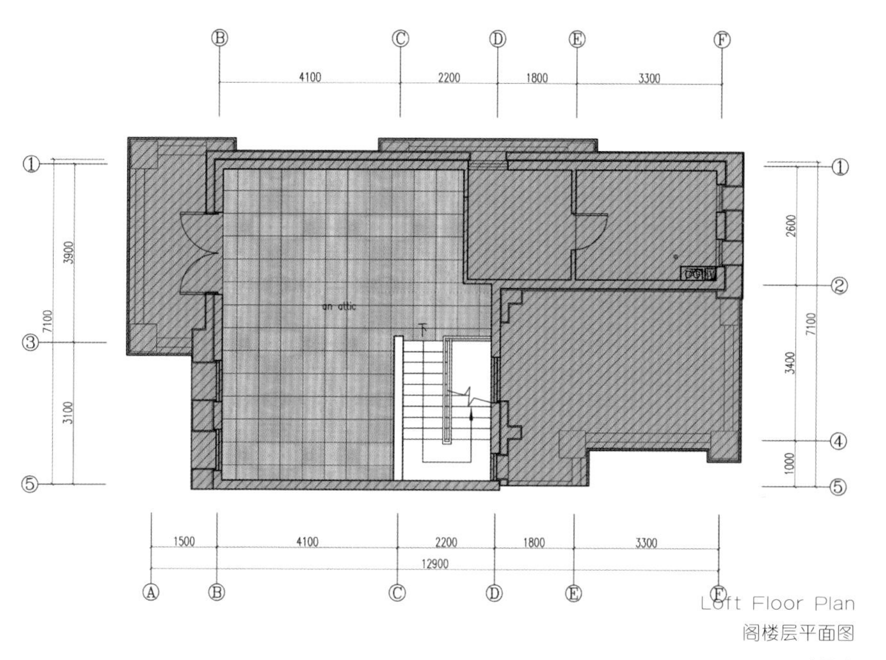

Loft Floor Plan
阁楼层平面图

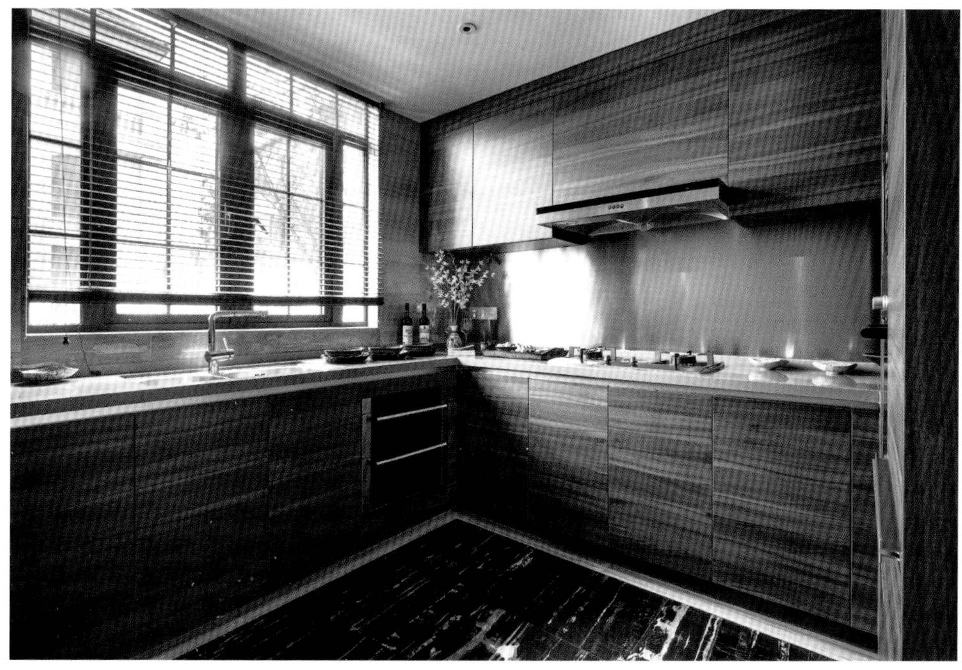

项目是东西方文化碰撞的产物，在尊重东方传统生活方式的基础上，融入西方艺术文化，营造新亚洲风格，将主题生活方式展现出来。私密空间温馨，注重享受。工艺给人新的视觉冲击和新的思维冲击（主题式展现）。

项目布局上把空间作为一个整体系统，这个系统以人为中心，包括天地万物。环境中的每一个整体系统都是相互联系、相互制约、相互依存、相互对立、相互转化的要素。功能布局就是宏观地把握各子系统之间的关系，优化结构，寻求最佳组合。

项目以简洁、明晰的线条和优雅、得体的装饰，展现出空间中华美、富丽的气氛，表达了一种随意舒适的风格，将家变成释放压力、缓解疲劳的地方，给人以典雅宁静又不失庄重的感官享受。简洁的天花设计和自然木纹形成几何图案的地板，把握了美式风格的简洁、对称、幽雅的精髓，空间中白色大理石立柱的运用则表达了一种更加理性、平衡、追求自由、崇尚创新的精神。

富丽的窗帘帷幔和水晶吊灯的搭配为以硬朗线条为主的空间中增添了一分柔美浪漫的气氛；而主卧及多功能厅的天花花纹仿佛天空中的云卷云舒，让理性的平衡中多了一丝丝的神秘。

宽大的沙发和椅子透着古朴与沉稳。坐在其中，不管是惬意的阅读抑或是沉醉于醇香的红酒都有让人从五星级酒店或者是办公室中解脱，回归到家中的舒适和随意。湖蓝色和桃红色装饰品的运用为深褐色的沉稳增添了一抹亮色，客厅和餐厅则在锡铅合金烛台、雕花边柜的装饰中，呈现出独特的韵味。

Green Lark Gede Center Hotel A1 Villa

绿湖 · 歌德廷中央酒店 A1 别墅

Designer: Simon Chong
Interior Design: Simon Chong Design Consultants Limited
Location: Nanchang, Jiangxi, China
Area: 680 m^2

设计师：郑树芬
室内设计：SCD香港郑树芬设计事务所
项目地点：中国江西省南昌市
面积：680 m^2

Keywords 关键词

Ultimate Luxury 极致奢华

Combination of Chinese and Western Elements 中西合璧

Simple and Elegant 淡雅别致

Furnishings/Materials 软装 / 材料

Marble Mosaic, Marble, Floor Tiles, Hollowed-out Screen, In-floor Heating System

云石马赛克、大理石、地砖、镂空屏风、地暖系统

By adopting colors and furnishings to achieve the division of function zones, different styles with the changes of color, lights and the atmosphere have made each floor unique, and the designers have brought out the concept of "using colors to express the space, adopting perception to enjoy the feelings".

With the spacious area of over 240 m^2, the basement is designed with bar, billiards room, game room, as well as two British style bathrooms. The semi-perspective architectural design makes the space brighter and gets better ventilation, while by adopting perfect and changeable colors to reveal the quietness and uniqueness of the basement. The capacious activity areas can hold 20~30 guests in the same time, as a space where people can communicate with each other freely and elegantly, which have showed the cultured characters and hospitality of the master.

The designs for the ground floor are with unique creativity. The main activity area for this floor is the living room, and the hallway decorated with the black and brown hollowed-out screen with rich Chinese elements, shows out the dignity and elegance of the master. The living room is designed with lots of different elements. The classical furniture is in the modern style. The rare fabric, velvet, droplight, stovepipe, paintings as well as the customized floor tiles imported from Italy, all of them have emphasized the center point of the living room, and the ultimate luxury. Especially for the curtain which contains cotton, linen, silk and so on, is with different reflection, hence different colors and effects are caused when the sunshine gets in.

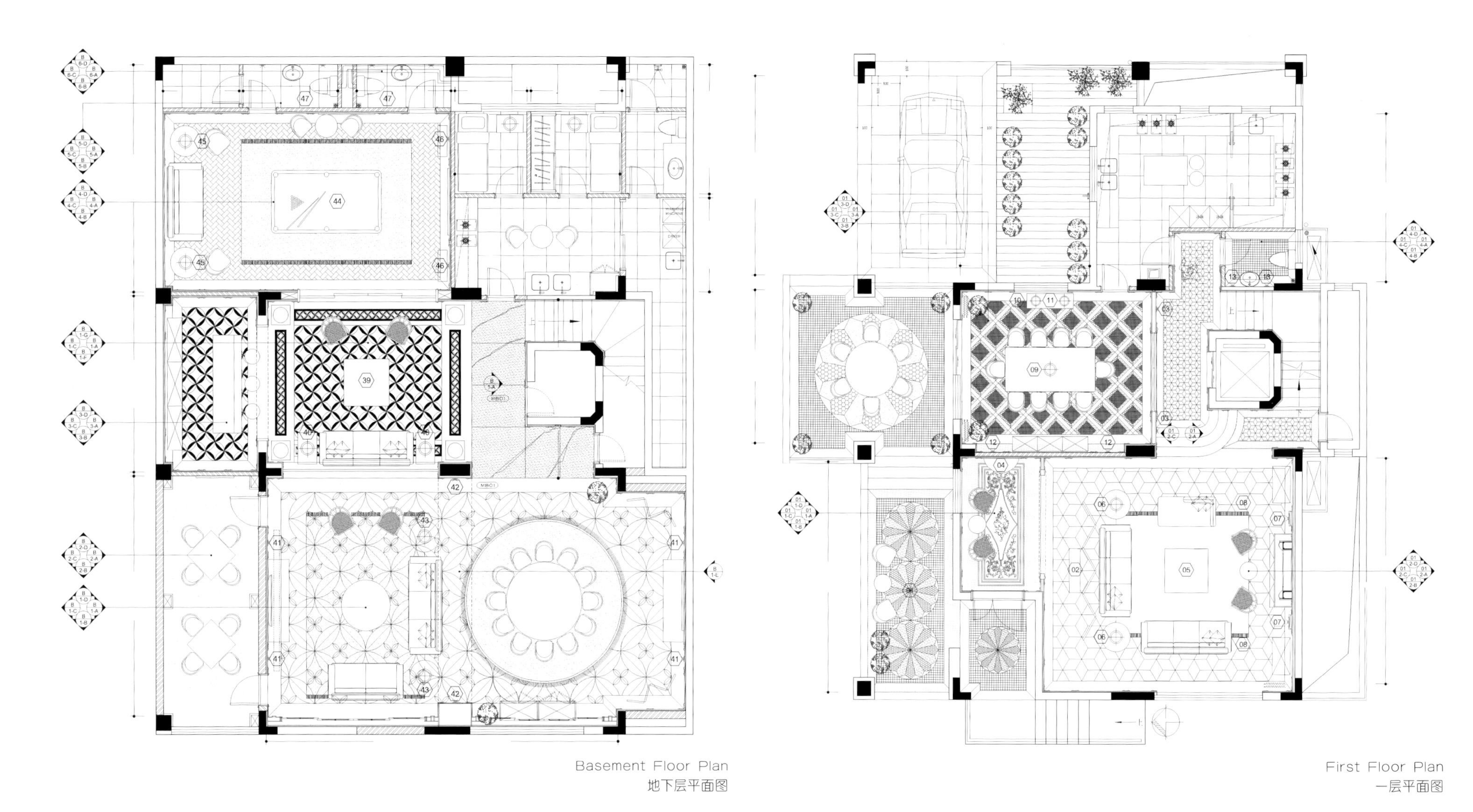

Basement Floor Plan
地下层平面图

First Floor Plan
一层平面图

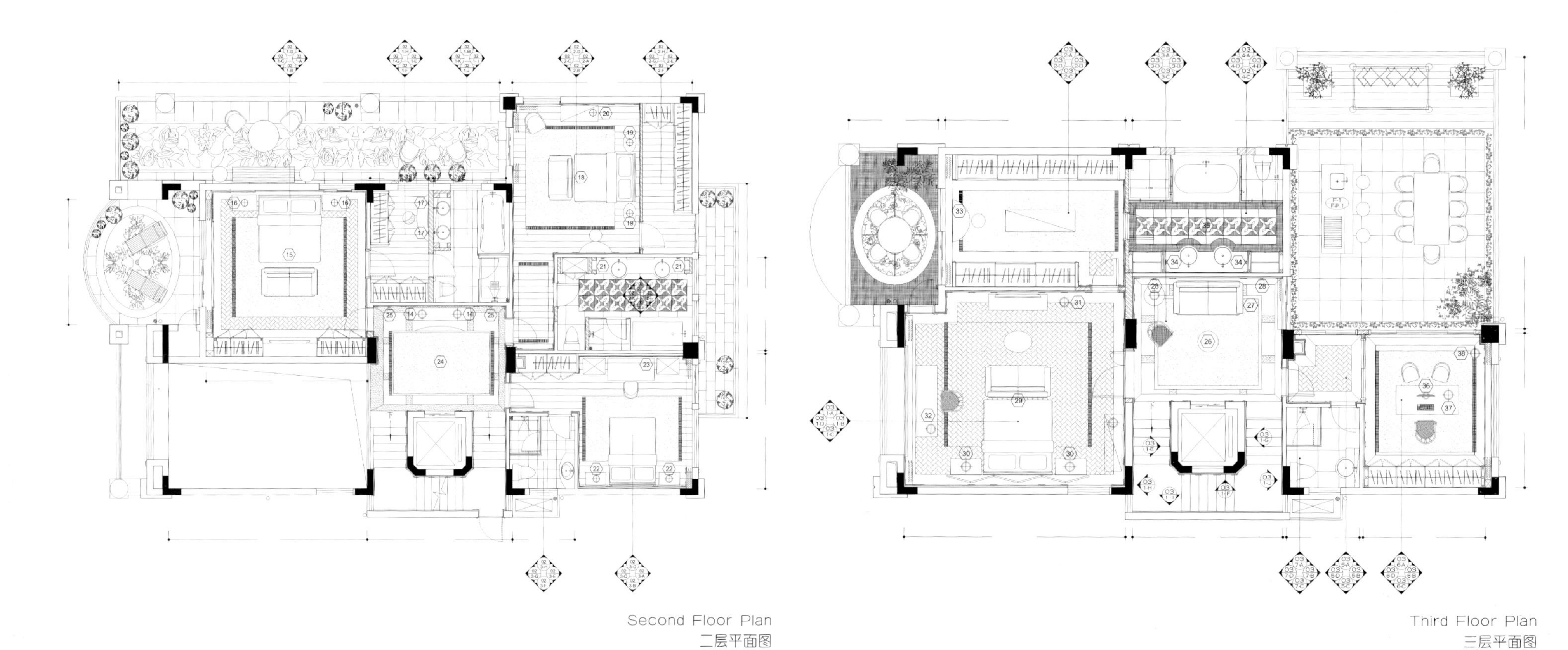

Second Floor Plan
二层平面图

Third Floor Plan
三层平面图

The second floor is the bedrooms for the parents and children of the master. Specialized colors are used for the bedrooms, the boy's bedroom is in yellow to express the elegance and teach him to open the heart, while the girl's is in purple which is cute and charming, as well as graceful.

The master bedroom is on the third floor. Simple and elegant designs are chose for this space. The cooperation between the colors and lights, as well as the patterns combining with Chinese and western style have expressed the modern taste in these classical elements and integrated the gorgeous and concise style together. Besides, the circled dome and customer-sized Italian marble mosaic have showed the luxury and connotation of the space to the maximum.

设计师利用色彩和布置来实现空间功用的区分，不同的空间都拥有不同的风格，颜色、灯光、气氛都有所变化，每一层都有独特的作用，实现了“用色彩表达空间，让感观享受感觉”。

地下一层面积约240 m^2。内部空间大，因此设计师布置了酒吧区、台球室、棋牌游戏室以及两个英式风格的独立洗手间。半透视的建筑设计显得明亮而通风，而色彩的运用和变换，让人能享受到地下空间的宁静和独特性。活动空间充裕，可招待20 ~ 30个客人，实现了“轻松沟通、高雅交流”的场所，展示了主人儒雅的个性和豪爽热情。

一楼空间设计布局体现独到的创意水平。这一层以客厅为主要活动场所，进门就是玄关。玄关处有一个充满浓郁中国元素的镂空屏风，黑中带棕的色彩，既显示尊贵又典雅。客厅也使用了众多元素的混搭。家具在古典中透着现代味道。空间使用名贵布料、丝绒、吊灯、火炉、壁画以及从意大利定制的进口地砖，不仅突出了中心点，而且体现了一种极致的奢华。特别是客厅的窗帘，用料中包括棉、麻、丝等多种面料，各种面料反光程度不一样，在阳光下呈现出色彩跳动和变化的效果。

二楼为父母与孩子的居处。居处在颜色使用上独具匠心，设计师特地将男孩的卧室调为黄色，可育养贵气，扩心胸；而主色调为紫色的女孩卧室，不仅娇俏，也增添了几分雅致。

三楼为主人翁居所。居所颜色淡雅别致，尤其注重色彩与光影的结合，中西结合的各色图案，古典中透露着时尚，华丽与简约的风格巧妙融为一体，配合圆形穹顶和特别定制的意大利云石马赛克，充分将主人空间的华贵、内涵诠释得淋漓尽致。

New Chinese Style

新中式风格

Graceful and Exquisite
精炼质朴

Traditional Flavor
古色古香

Understated Elegance
含蓄秀美

Vanke Cheerful Bay Duplex House A2

重庆万科悦湾 A2 洋房复式

Interior Design: Matrix Interior Design
Location: Shapingba district, Chongqing, China
Area: 217 m^2

室内设计：矩阵纵横设计团队
项目地点：中国重庆市沙坪坝区
面积：217 m^2

Keywords 关键词

Chinese Classical Element 中国古典元素

Elegant & Simple 素雅古朴

Lofty Style 格调高雅

Furnishings/Materials 软装 / 材料

Stone, Rosy Golden Stainless Steel, Leather Hard Pack, Coated Glass, Shell Mosaic, Imported Wallpaper, Baking Sheet Finishes

石材、玫瑰金不锈钢、皮革硬包、镀膜玻璃、贝壳马赛克、进口墙纸、烤漆板饰面

The interior, mainly in colors of black, white and gray, is to create a quiet and relaxing atmosphere. In order to match with the main tone, the designer uses soft and moderate intensity lights in the interior. Meanwhile, the designer uses French window in the living room, which directly leads the exterior lighting and beautiful natural scenery into the interior, achieving the purpose of sight seeing and adding natural ambiance of the interior. The application of bright lines and space configuration highlights the unique skill of the designer. The carpet with grid lines in the living room maintains and embellishes the neat space. The project is just like an elegant and stretched ink painting, which is gracious and simple, solemn and introverted in a lofty style. In the details, it advocates natural interests, reduces excessive color decorations and integrates Chinese classical elements. The traditional furniture, porcelains, antiques and aromas inadvertently build a simple and elegant space, transmit the profound implication of the traditional culture and express people's pursuit of a tranquil life.

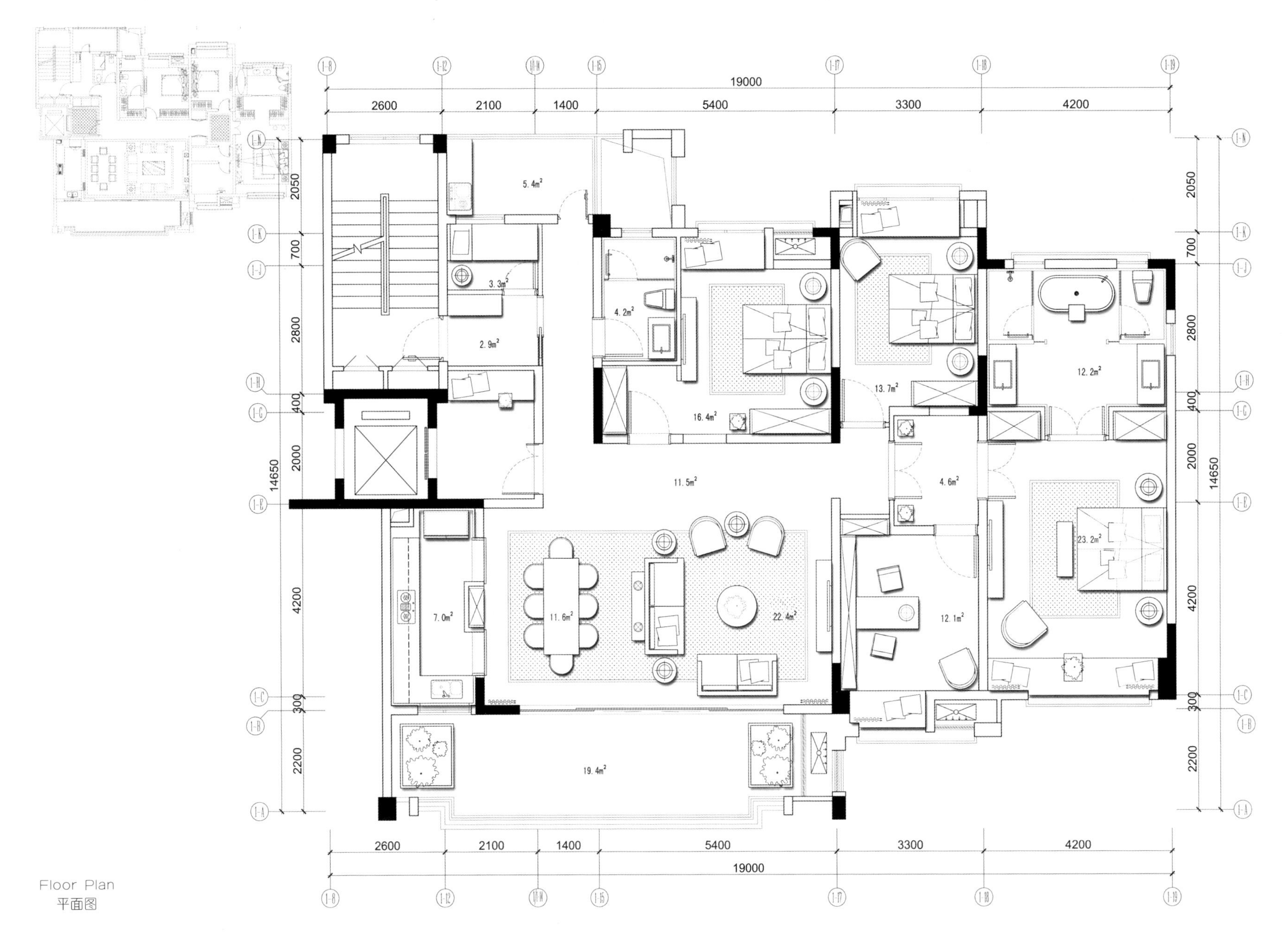

Floor Plan
平面图

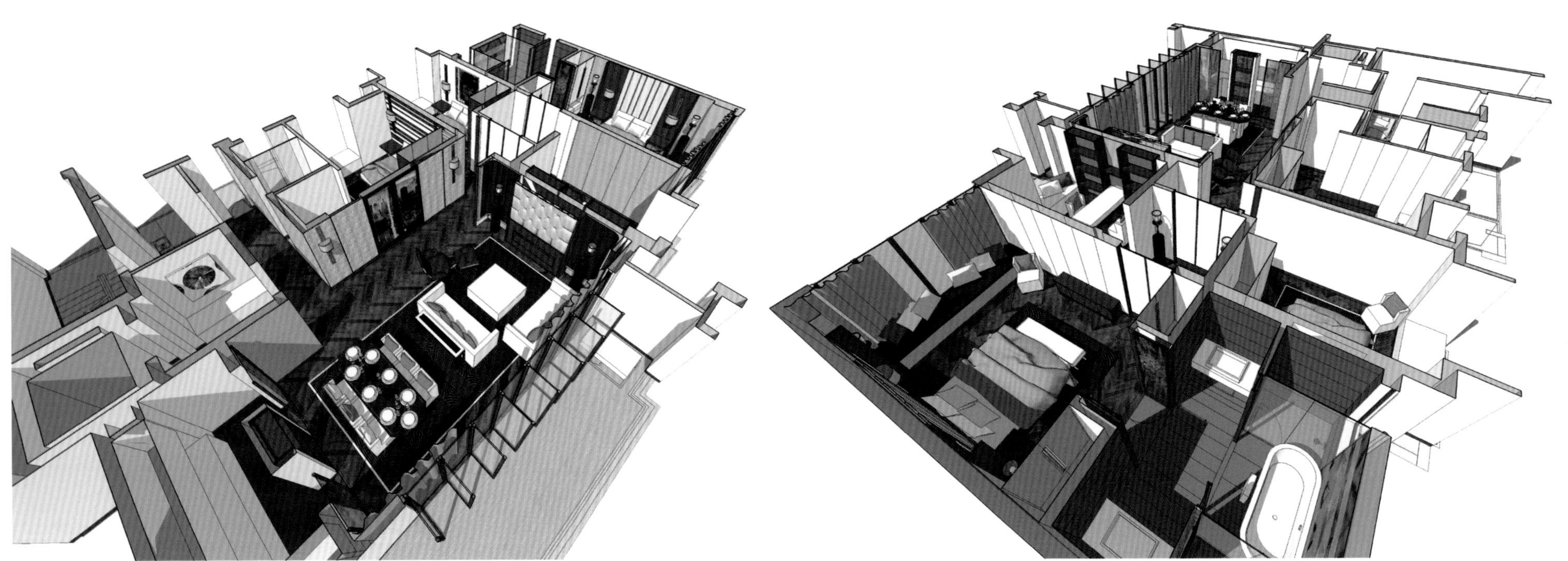

项目以黑、白、灰为主色调，营造出一种静谧而让人放松的氛围。为了配合主色调，设计师在室内的灯光应用上以柔和和亮度适中的灯光进行搭配。同时，设计师在客厅位置设计了落地玻璃窗，直接将户外光线和悠然美景引入室内，既达到欣赏美景的目的，又增加了室内的自然气息。而明快的线条和明朗的空间构图的运用亦彰显出设计师手法娴熟独到。客厅的网格线地毯不仅保持着空间的规整，而且起到点缀的作用。项目整体犹如一幅飘逸舒展的水墨写意画，清雅质朴、格调高雅、沉重内敛。在装饰细节上崇尚自然情趣，剔除了过多色彩装饰，融入了丰富的中国古典元素。传统家具、瓷器、古玩香薰等在不经意间打造了素雅古朴的至美空间，传达着传统文化的深远意蕴，表达人们对宁静致远的生活境界的追求。

GUND

Metropolitan · Guiyang Chinese Monochromes

贵阳中航城 · 水墨画

Designer: Simon Chong, Amy Du
Interior Design: Simon Chong Design Consultants Ltd.
Location: Guiyang, Guizhou, China
Area: 230 m^2

设计师：郑树芬、杜恒
室内设计：SCD香港郑树芬设计事务所
项目地点：中国贵州省贵阳市
面积：230 m^2

Keywords 关键词

Chinese Charm 中式韵味

Relaxing & Fashionable 轻松时尚

Chinese Monochromes Imagery 水墨画意境

Furnishings/Materials 软装 / 材料

Wooden Grid, Chinese Chair, Butterfly Orchid, Tea Set, Auspicious Beast Pottery, Landscape Painting, Hanging Photograph, Lotus Lamp, Enclosure-pattern Fabrics

木质格栅、中式单椅、蝴蝶兰、茶具、瑞兽陶器、山水画、摄影挂画、莲花灯、回型布艺

The project is a three floored duplex. It has a basement and a garage due to good house type structure. The basement includes tea room, study room, audiovisual room and maid's room, while the ground floor consists of living room, dinning room, main bedroom, elderly's room, children's room and study room. The hallway of the ground floor is a highlight of this house. First came into sight is a painting with abstract imagery. Meanwhile, the higher bird sculpture and the lower plum flower are well scattered, manifesting the modern Chinese beauty.

The wooden grid ornament on the wall of the living room is a great feature. The Chinese chair, longevous pine on the long table, butterfly orchid on the tea table, tea set and auspicious beast pottery jointly manifest the imagery of Chinese painting. While the sofa are in concise fashion to show the beauty of modern form. The throw pillows are in red for proper embellishment, and the stretched plum flowers on the throw pillow express the "Chinese Red" vividly and completely. The whole space is integrated with both Chinese and modern flavor, making people feel like being with the natural landscape.

Similarly, the dinning room is in modern manifestation. By making the tables and chairs with enclosed-pattern fabrics as the leading features, the designer skillfully integrates various eastern elements into the design. The embellishment of red table linen and flower decorations on the table, though modern and fashionable, is also a detailed manifestation of "Chinese Red", which shows the perfect combination of tradition and modern.

The main bedroom is in gray tone, while the bedding is in light purplish gray to make the whole space soft and warm. The bird painting above the bed makes the life tranquil. Meanwhile, the lotus lamps at the bedside show the Chinese charm and further extend the imagery of Chinese painting in the living room. Besides, the black and white photograph hanging in the main bedroom makes the whole space relaxing and fashionable.

项目户型为复式楼，共有三层。由于户型结构良好，项目拥有负一层和车库层。负一层为茶室、书房、视听室及工人房；一层为客厅、餐厅、主卧、老人房、儿童房及书房。一层的玄关处是项目户型的亮点之一。首先映入眼帘的是一幅抽象意境的画作。同时，小鸟雕塑与梅花一高一低，错落有致，彰显现代中式之美。

客厅墙面以木质格栅装饰为亮点。中式单椅、几案上的延年松、茶几上的蝴蝶兰、茶具及瑞兽陶器共同演绎出水墨画的意境。而沙发又以简约时尚来体现现代的形式美。沙发上的抱枕以红色作恰如其分的点缀；抱枕上的舒展着的梅花将“中国红”表达得淋漓尽致。整个空间将中式与现代相互交融，让人置身山水之间。

同样，餐厅以现代中式为表现形式。以回型布艺的餐椅为主角，设计师将各种东方格调元素微妙地融入其中。餐桌上红色餐布及花艺的点缀，虽然现代时尚，但也是“中国红”的细致演绎，体现了传统与现代的完美

结合。

主卧以灰色为主色调，而床品则以淡紫灰色呈现，让整个空间显得柔和温馨。床头的一副小鸟图让生活变得恬静。同时，床头的莲花灯表现着中式的韵味，更是延续了客厅“水墨画”的意境。另外，主卧的黑白摄影图挂画让整个空间显得轻松而时尚。

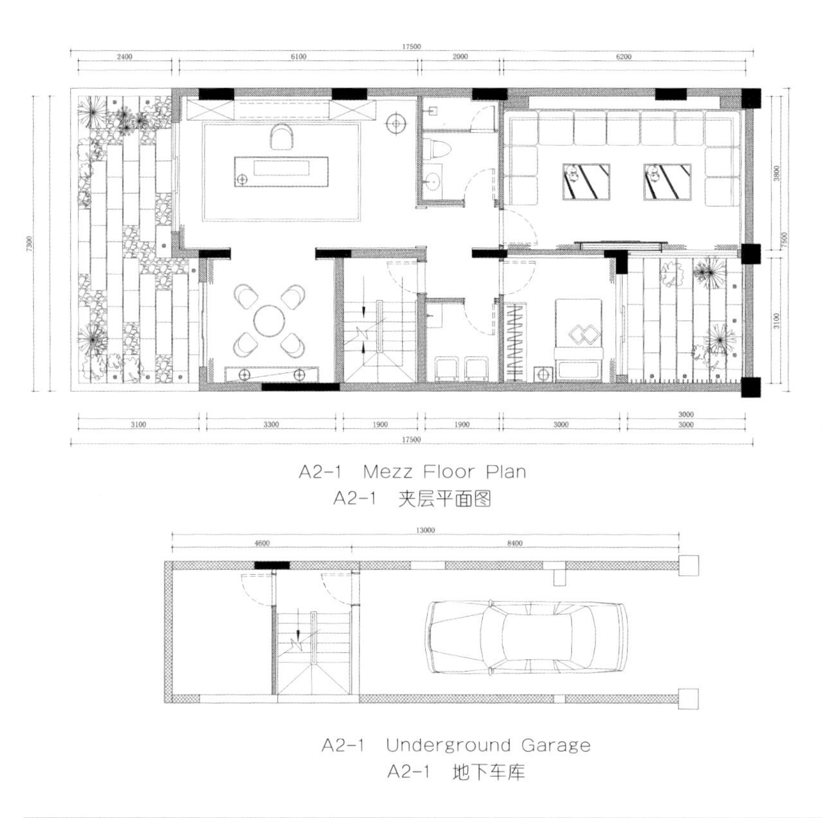

A2-1 Mezz Floor Plan
A2-1 夹层平面图

A2-1 Underground Garage
A2-1 地下车库

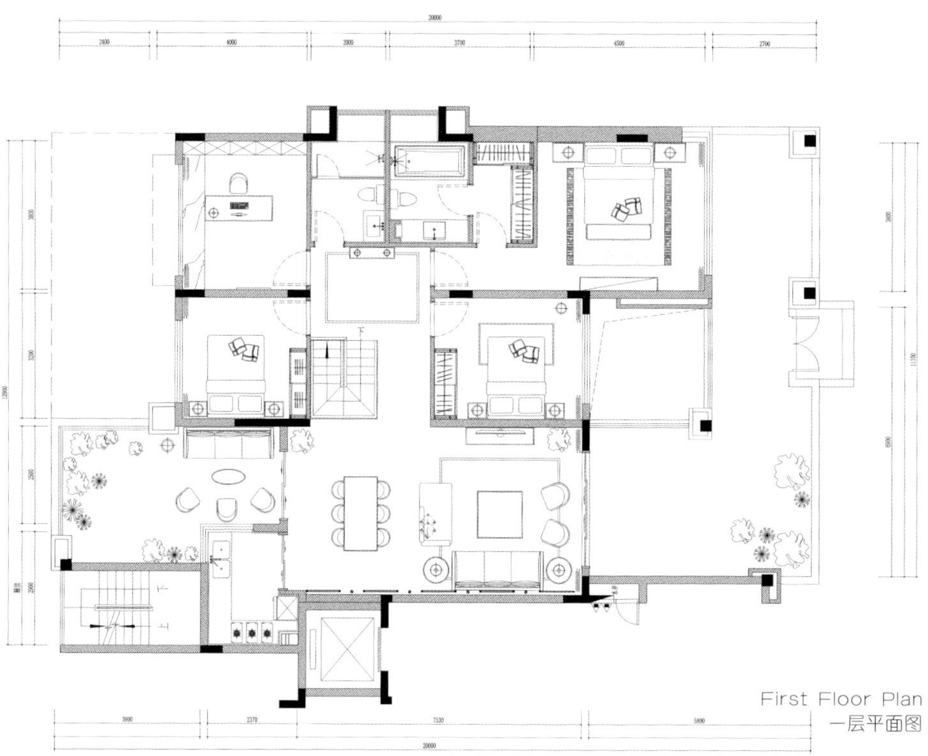

First Floor Plan
一层平面图

Yu Villa
于舍

Designer: Xu Jianguo
Interior Design: XJG Architectural Interior Design Co., Ltd.
Location: Hefei, Anhui, China
Area: 480 m^2

设计师：许建国
室内设计：合肥许建国建筑室内装饰设计有限公司
项目地点：中国安徽省合肥市
面积：480 m^2

Keywords 关键词

Return to Nature 返璞归真

Oriental Charm 东方神韵

Elegant Style 格调清雅

Furnishings/Materials 软装 / 材料

Log, Stone, White Brick, Ash Wood Veneer

原木、石材、小白砖、水曲柳木饰面

Considering the geographical environment, personal character and oriental beauty, the designers carefully designed door and window through an elaborate consideration and planning, using a large number of the warmest and the most emotional wooden elements and natural materials, to create a space full of natural flavor and human interest. In view of the householder's three generations under one roof, so it is clear in space division. The first floor of public space advocates humanistic feelings; the second floor includes rooms for the elderly and guests, emphasizing functional convenience; the third floor is the space for master room, paying attention to integration; the fourth floor takes daughter's study abroad experience into account, thus combines the Chinese and French style perfectly.

Logs and stones are natural materials. Their shapes, varying from light changes, become soft and full of vitality and show oriental charm of innocence, tranquility and nature. Project's main color is pure wood which highlights elegant, comfortable style and well-arranged pattern of hierarchy, fully embodies the harmonious dialogue between man and nature, and fully expresses leisure, comfortable and natural life.

Designers express the beauty of simplicity straight from nature. Detaching from the surface form of art and tasting the beauty of mystery, thus they keep the project far away from the city noise and return to simple, comfortable and quiet life.

项目设计师从地域环境、人物性格、东方之美出发，通过精细的考量和规划，采用大量的最优温度、最有感情的木质元素和天然材质，对门和窗的精心设计，力图打造出一个充满自然气息和人情味的空间。考虑到户主三代同堂，所以在空间划分上也是精雕细琢。一层公共空间倡导人文情怀；二层是老人房及客房，注重功能的便捷；三层是主人房空间，注重一体化；四楼女儿房则考虑到户主女儿的留学经历，融合法式风格，中西的完美切合。

原木和原石是自然生来自有的材料。它们随光线变化而变化的条形，柔和且富有生命力，兼具东方之神韵，纯真、宁静、自然。纯净木色为项目主色调，突出格调清雅惬意，错落有致的格局层次，充分体现人与自然的和谐对话，充分表现悠闲、舒畅、自然的生活情趣。

设计师直取本质，表达朴素之美，从表面的艺术形式超脱出来，品味幽玄之美，从而远离都市喧嚣，让生活回归质朴、舒适和宁静。

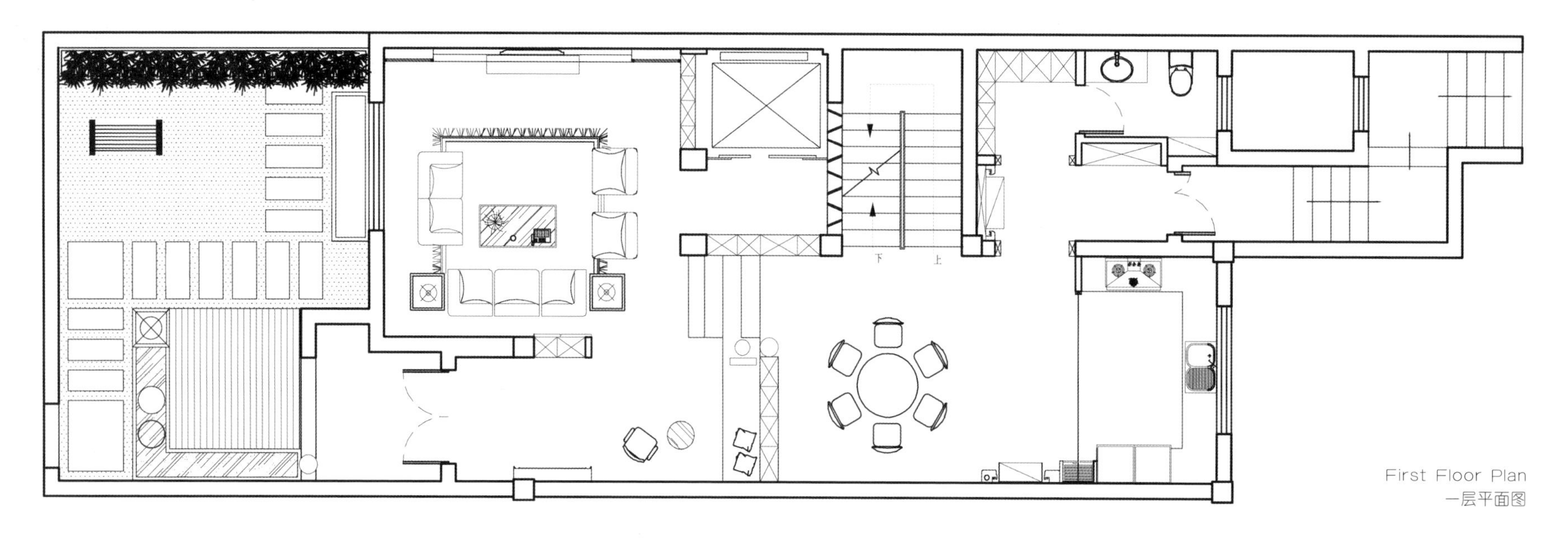

First Floor Plan
一层平面图

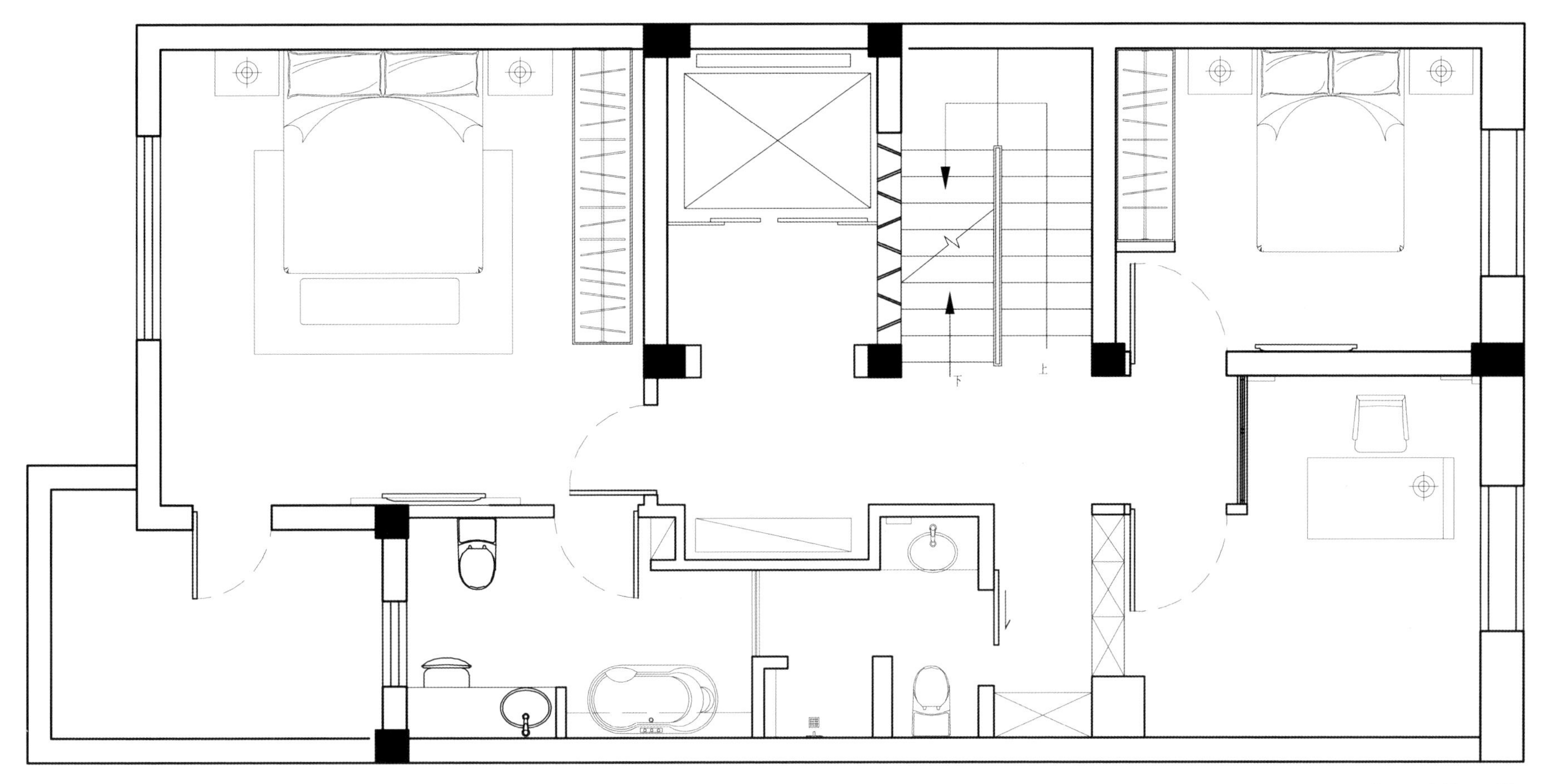

Second Floor Plan
二层平面图

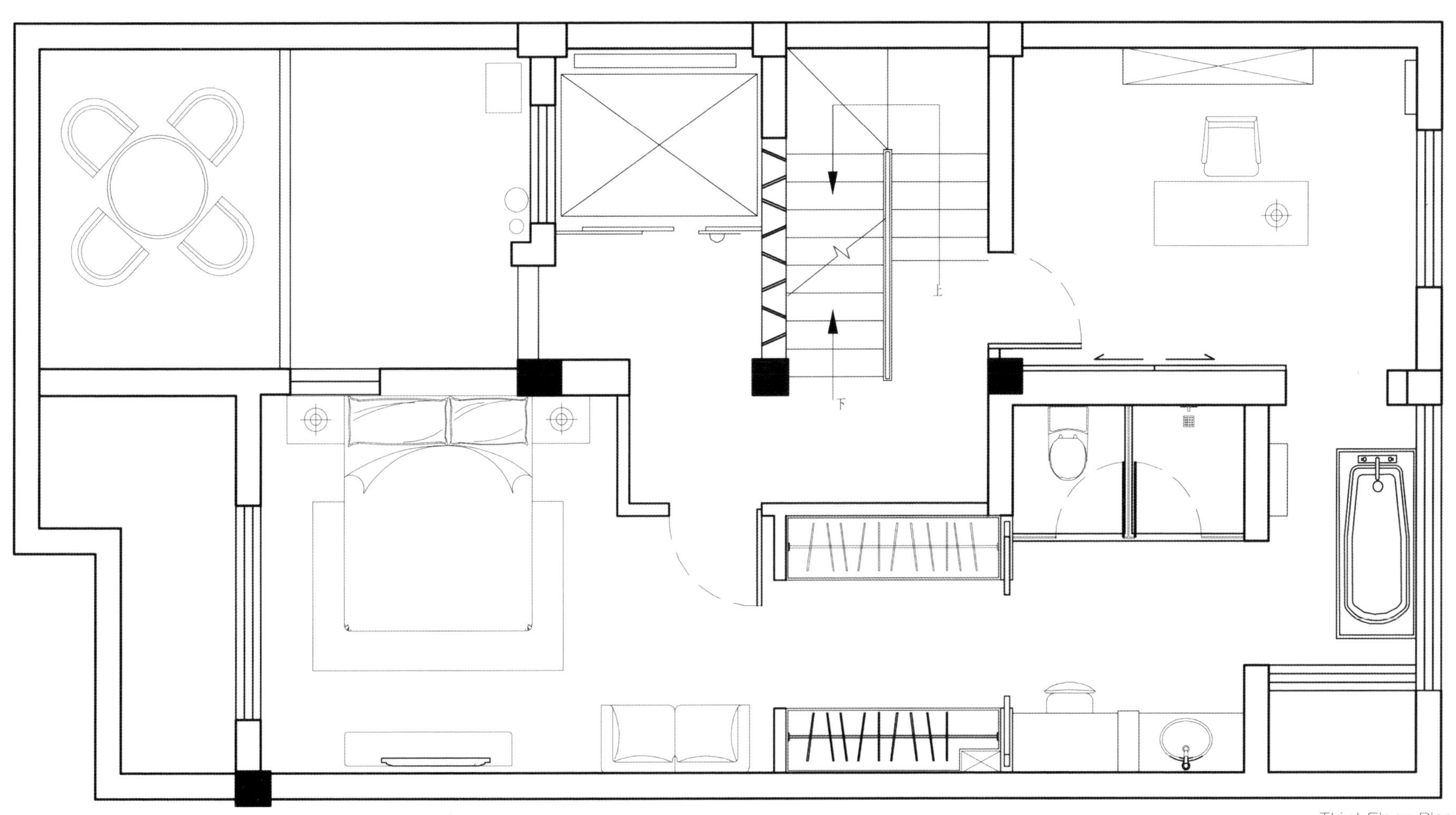

Third Floor Plan
三层平面图

燕雀歡歌桃樹紅風箏飛
舞入晴空乳兒坪上蹣跚
步過客忽知春意濃
甲午春
子娜書

Modern Style

现代风格

Fashionable and Gorgeous
时尚华丽

Lively and Innovative
活泼创新

Simple and Functional
简洁实用

Zhongshan Xiuli Lake Building 8 C type – Villa Sample House

中山秀丽湖项目 8 幢 C 型—别墅样板房

Designer: AJAX LAW
Interior Design: One Plus
Location: Wuguishan District, Zhongshan, Guangdong, China

设计师：AJAX LAW
室内设计：壹正企划有限公司
项目地点：中国广东省中山市五桂山区

Keywords 关键词

Stillness in the Midst of Motion 动中有静

The Theme of Books 书籍主题

Meticulous and Personalized 细致个性

Furnishings/Materials 软装 / 材料

Stainless Steel, Wood, Carpet, Spotlight

不锈钢、木材、地毯、射灯

This project is a villa sample house with the design theme of "Books". Through the static state and dynamic state of "book" and its own elements designed by designers, the individual impression of this villa is enhanced.

In the living room, the designer designed the feature wall as a book that can be read page by page. Fine cutting woods are made to be like a slope when turning book, full of dynamic sense. In an unbroken line, books are in overlapped display, showing a slant height curve, which let people strongly feel about dynamic state of book, amazing creativity of the designers, the infinite possibilities of using different materials and the meticulous and personalized effect of wood.

Wine chest also echoes with the design of feature wall in living room. Inclined effect appeared upon opening the books is designed into a grid storage racks, with wine bottle placed therein, just like the wine are fixed in each page of a book. The bar counter is also made of exquisite strip, showing each page in book.

The room reveals the "stillness in the midst of motion" of book. Master room, bedroom 1, bedroom 2, family cinema and the pattern design of carpet in living room are designed to be books in symmetry to the chaos. Soft furnishing is also specially added books, strengthening the feeling of integrity.

The back of bed in master room is deliberately designed into the state of sheets of stacked papers, in addition to show the static station of "book", strengthening the style and taste that the master room should have.

On the other hand, staircase armrest uses stainless steel materials. Stainless steel blade with different height is neatly displayed. Cabinets and bathroom in elevator lobby are designed similarly, echoing each other.

Because the book is made of trees, the whole villa sample house uses varying degrees of brown color. Different regions show different forms of books, just like being surrounded in the wonderful world of book.

项目为别墅样板房，以“书”为设计主题，设计师分别透过“书”的动态、静态及其本身的元素作为展示，加深访客对此别墅样板房的独特印象。

位于客厅的特色墙，设计师将之设计成有如被一页页翻阅着的书本。运用精细切割的木材，做成像翻动中的书本一样的斜面，充满动感。连绵不绝的书本，重叠摆放般，展现出来的斜高曲线，不单让人强烈感受到书动态的一面，更让人对设计师的创意啧啧称奇，让人了解到运用不同物料所创造的无限可能性，感受到木材展现出的细致而有个性的效果。

酒柜同样呼应了客厅特色墙的设计，将打开的书本呈现出来的斜面效果设计成格子储存架，酒瓶放置其中，有如将酒固定在每一页书间。酒吧台同样以精致的板条做成，寓意为每一页书页。

房间则将动中有静的“书”的形态呈现出来，主人房、睡房1、睡房2、家庭影院及客厅地毯图案设计成乱中有序的书本，软装配饰上亦特别加入书的摆设，加强整体的感觉。

主人房的床背刻意设计成一张张纸重叠起来的形态，不但表现出“书”的静态，亦加强了主人房应有的格调和品位。另一方面，楼梯扶手运用不锈钢材料，用不同高度的不锈钢条整齐排放在一起，电梯、大堂的柜子、卫生间亦采用相近的设计，彼此呼应。

由于书是用树木造成，整间别墅样板房以不同程度的棕啡色为主色调。通过不同区域展现了书的不同形态，让人有如被书环抱，置身于书中的奇妙世界一样。

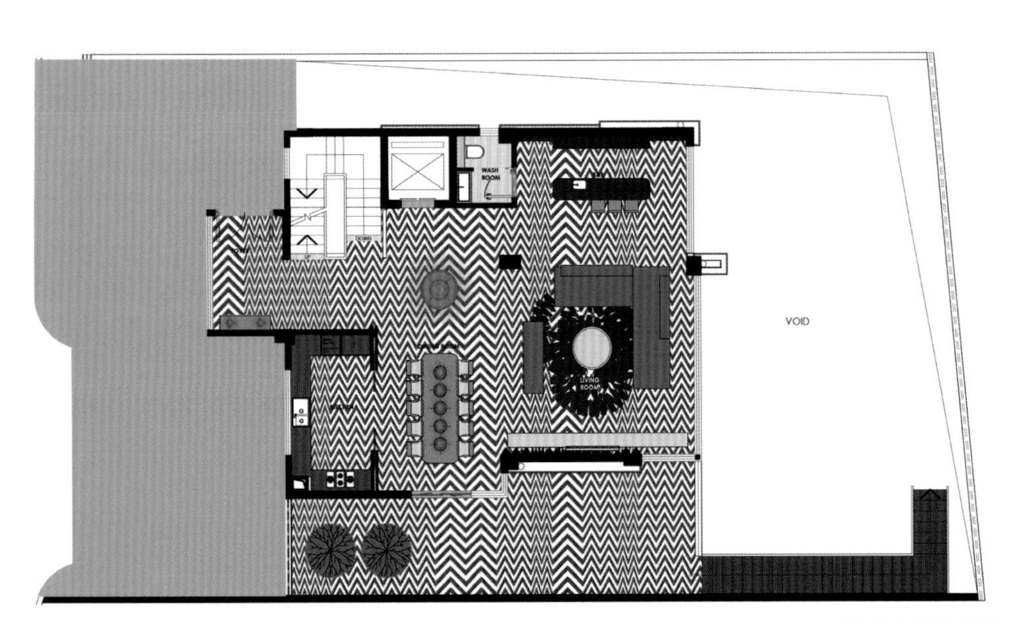

First Floor Plan
一层平面图

Second Floor Plan
二层平面图

Basement Floor Plan
地下室平面图

Third Floor Plan
三层平面图

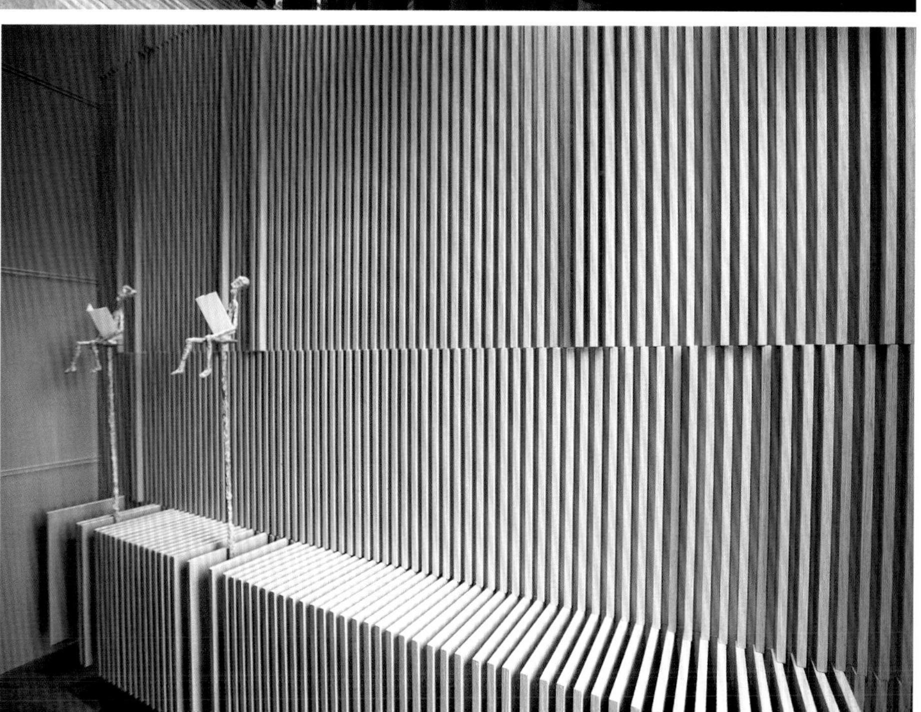

Logan Grand Riverside Bay, Villa Show Flat No. 61

龙光水悦龙湾 61# 独栋别墅样板房

Designer: Chen Kunming
Interior Design: Hover House Decoration Design Engineering Co., Ltd.
Location: Foshan, Guangdong, China
Area: 500 m^2

设计师：陈昆明
室内设计：深圳市世纪雅典居装饰设计工程有限公司
项目地点：中国广东省佛山市
面积：500 m^2

Keywords 关键词

Oriental Feelings 东方情怀

Modern and Fashion 现代时尚

Low-key and Luxury 低调奢华

Furnishings/Materials 软装 / 材料

Marble, Fabrics, Leather, Wallpaper, Wooden Facing, Plating Stainless Steel

大理石、布料、皮革、墙纸、木饰面、镀色不锈钢

With the elegant but low-key taste, the project has impressed people a lot. Simple colors and abundant sunshine in the unit have made the space more spacious. For example, the ornaments with elegant shapes hanging over the staircase have highlighted the grandness of the space. Meticulous designs can be seen everywhere in the villa, from the overall framework to the details, every single line and space is set on purpose to integrate into the charming living environment. In addition, the designs for the soft furnishings are also sophisticated, such as high-grade furniture both with the beauty of orient and modern times, luxury fabrics, graceful crystal lamps, artwork accessories and decorative lamps, all have expressed the beauty of the details. Elegant art taste of the master is also showed from the layout of projection booth, billiard table, bar, wine cabinet and so on. Rich layering are created by the colors, which have built up the high-end luxury living environment.

项目典雅与低调并重的气质令人印象深刻。别墅内部色彩简约朴实，光照充足，营造出宽敞的观感。以自然垂落于楼梯间的吊饰为例，其姿态灵动典雅，烘托出空间的利落与大气。无论是对整体框架的把握还是细节的处理都彰显出设计师的巧思，每一根线条，每一个空间都经过精心的设计和安排，似乎每一个元素都是为这个空间而生，完美地融合成了迷人的居住空间。设计师在软装上也独具匠心，质地上乘且融合东方情怀与现代时尚美感的家具配置、华贵的布艺、优雅的水晶灯、工艺品摆件、装饰画灯，都在大气中尽显细节之美。放映室、桌球台、吧台、酒柜等空间的处理，都体现了主人优雅的艺术品位。整个空间在选色上打造出了丰富的层次感，营造出低调奢华的高品质居室氛围。

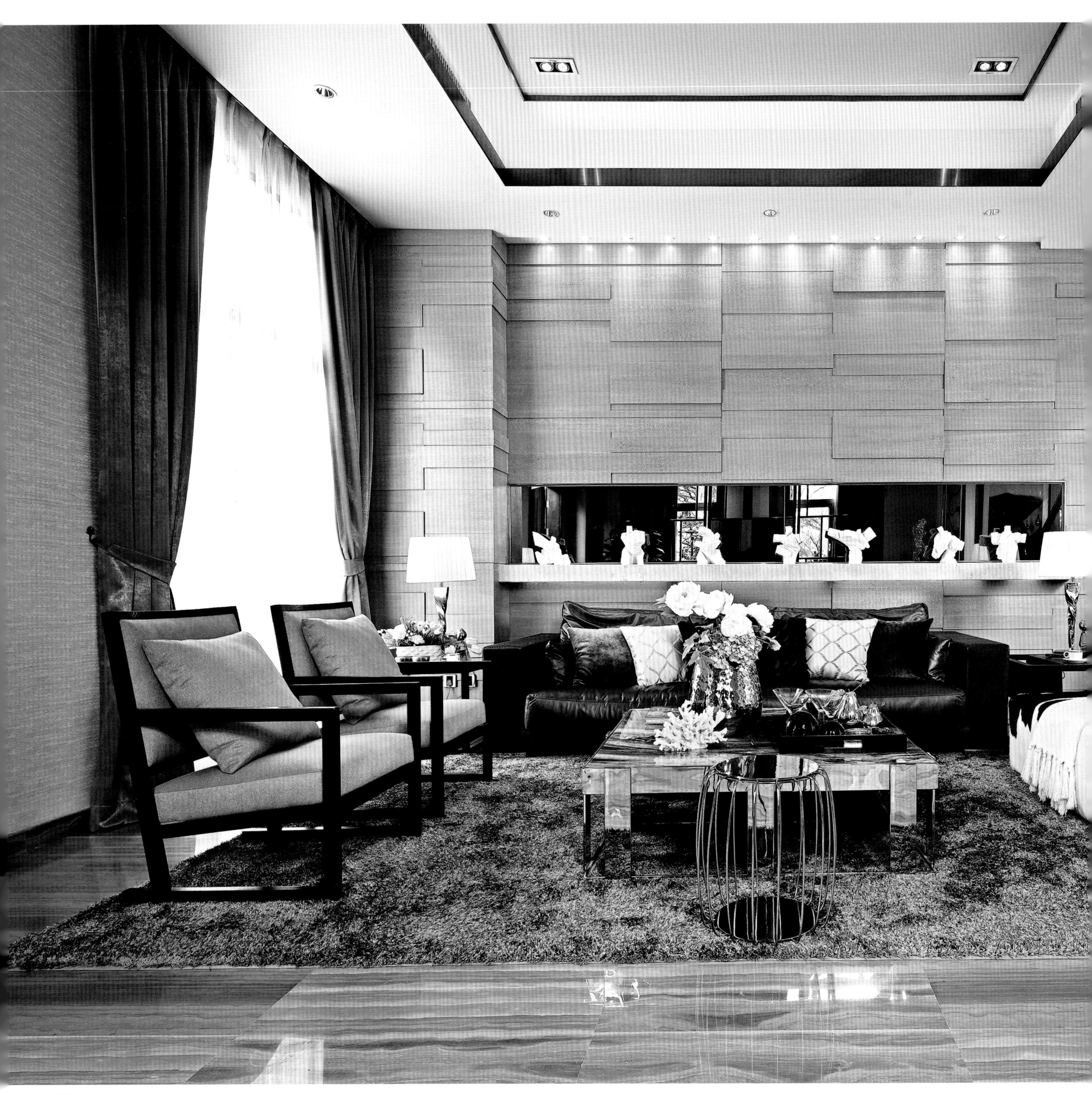

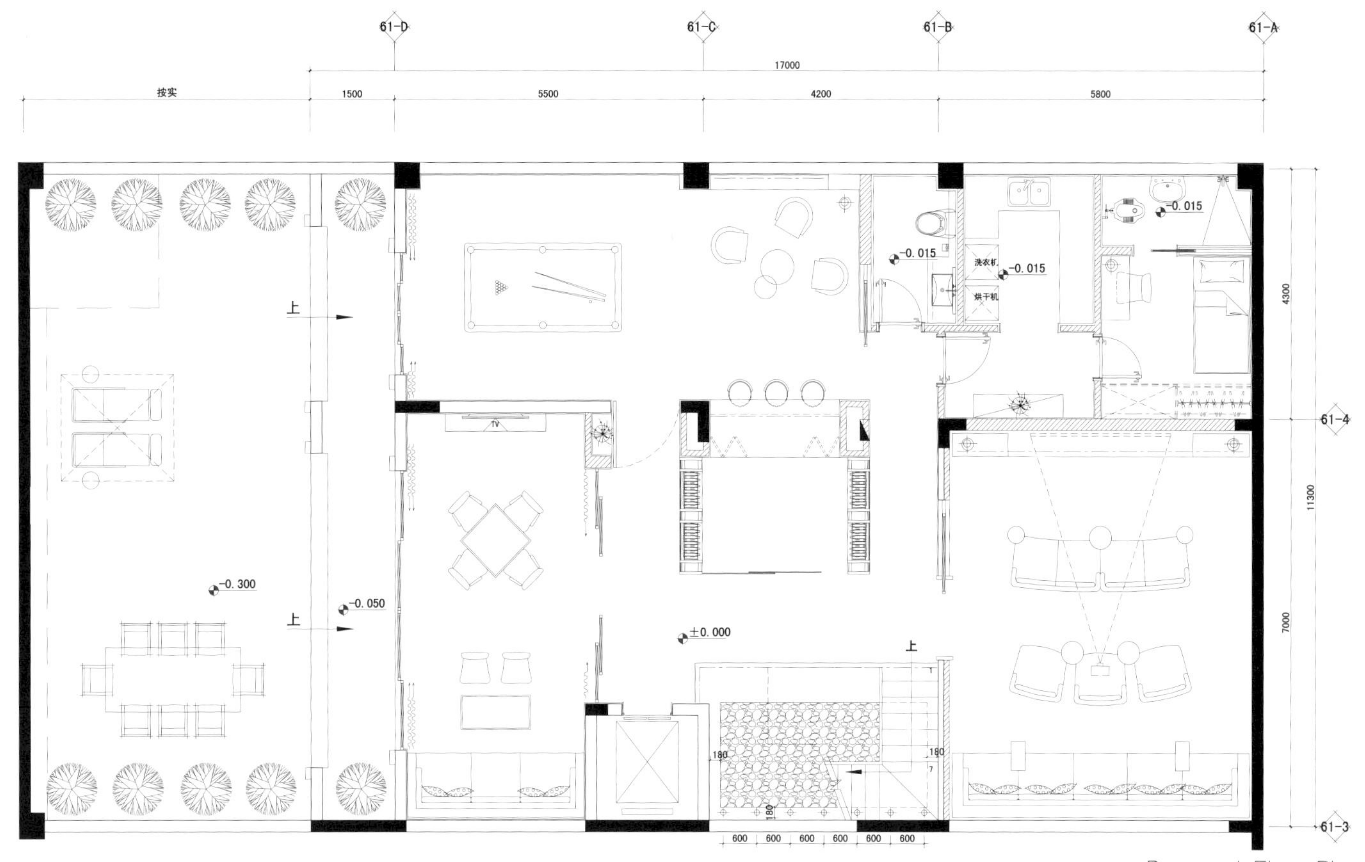

Basement Floor Plan
负一层平面布置图

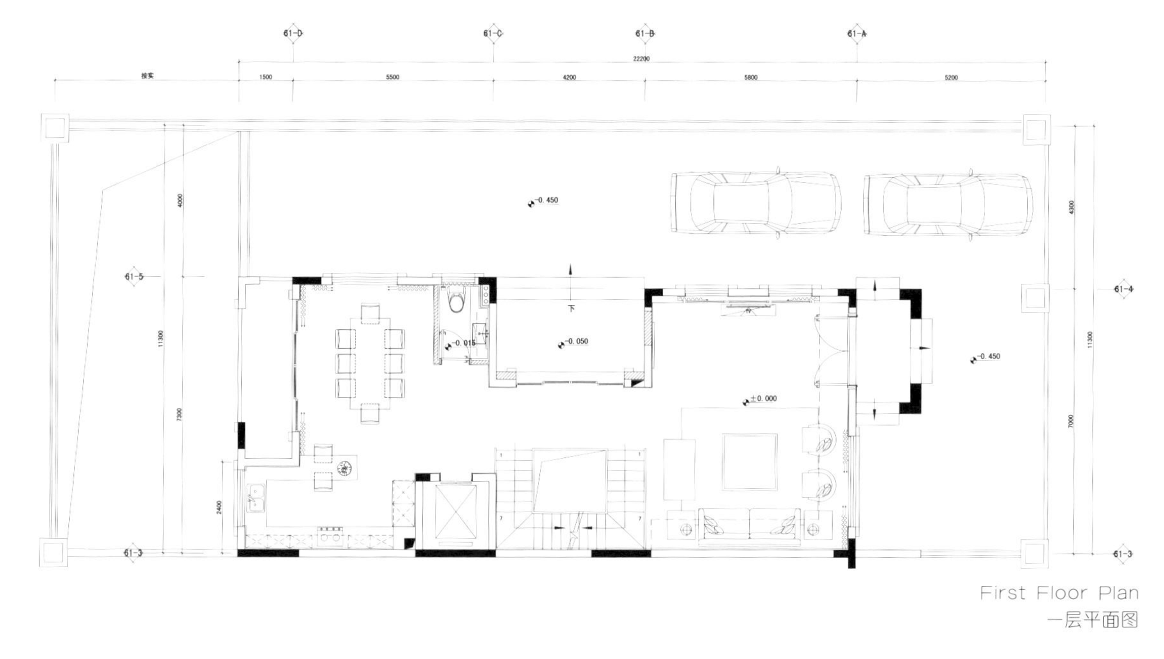

First Floor Plan
一层平面图

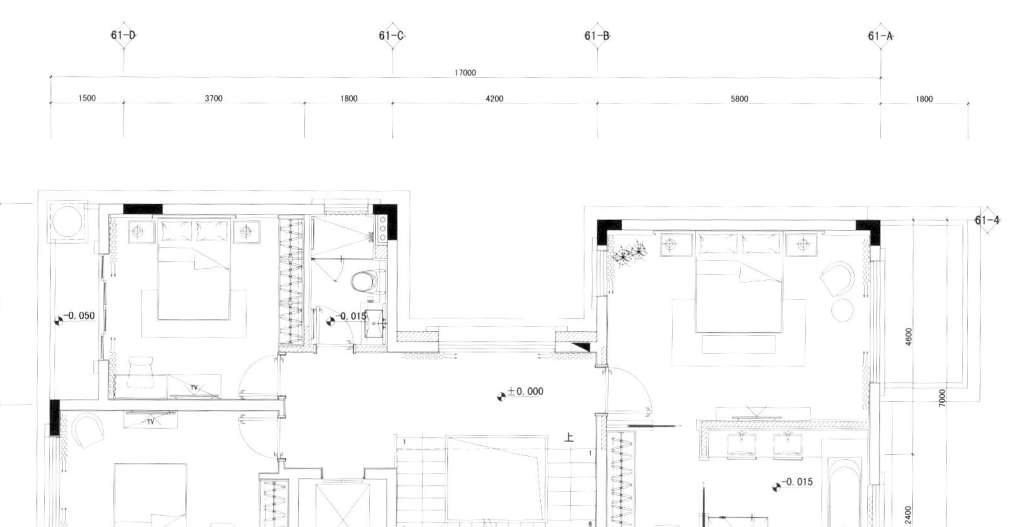

Second Floor Plan
二层平面图

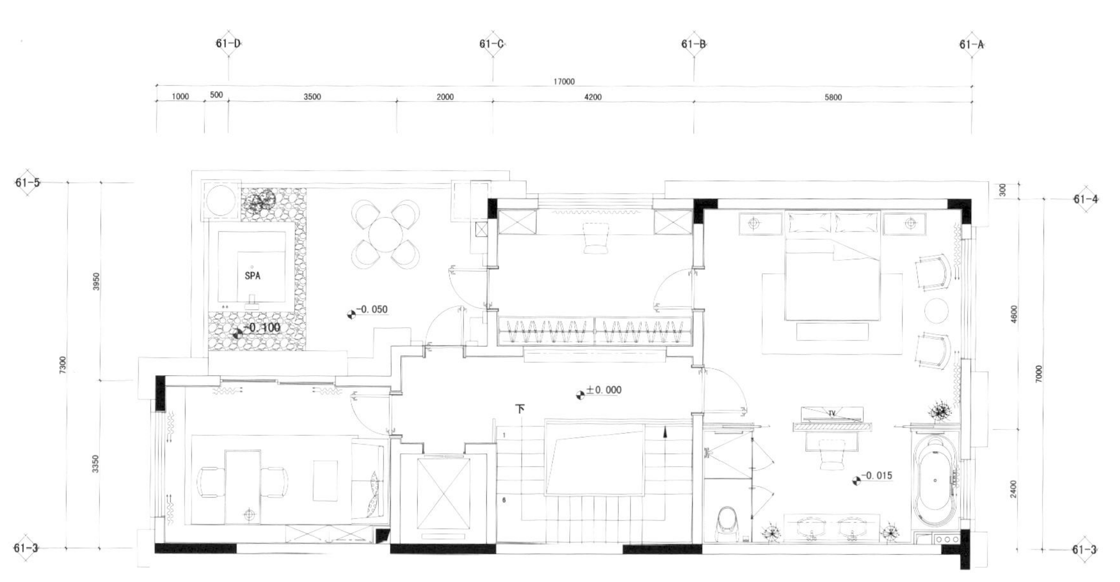

Third Floor Plan
三层平面图

MACALLAN
18
18

PRESLEY
HEPBURN

PRESLEY
HEPBURN

Under the Tree

树下

Designer: Shao Weiyan
Interior Design: CHU-studio
Location: Taoyuan County, Taiwan, China
Area: 314 m^2

设计师：邵唯晏
室内设计：竹工凡木设计研究室（CHU-studio）
项目地点：中国台湾桃园县
面积：314 m^2

Keywords 关键词

Light through Hollowed Ceiling 空中光井

Transparent Glass 清透玻璃

Sufficient Light 充足光线

Furnishings/Materials 软装 / 材料

Steel Structure, Carpentry, Polished Quartz Brick, Tempered Glass, Acrylic, Cold Paint, Carbonized wood

钢构、木作、抛光石英砖、强化玻璃、亚克力、冷烤漆、碳化木

The project uses growing big trees as the design imagery that corresponds to the scene of studying under the tree. This fully shows the householder's expectation of the children's growth and concern and care of the children's education. Moreover, except for embodying artistry, the big tree possesses the special function of showing "Education of Love".

The layout of the house is deep laterally and narrow in lateral, and the space would seem narrow when deducting the stairway and the rooms. The designer remains the original empty space and uses tempered glass, together with the tree branches imagery to create a facile and ventilated stair handrail, which can enhance the spaciousness and make the bright green the visual focus. Besides, the sufficient light through the empty space solves the insufficient light problem. The designer has specially created a hollowed ceiling view that varies with the nature.

The designer reduces the partition wall of each floor and the lighting & visual obstacles and remains the least amount of wall space partitions to maintain the openness, penetration and continuous entirety of the space. Meanwhile,it cuts the upper end of the partition wall, and installs transparent glass to achieve the effect of seeing the big tree at each place of the interior. It makes the lighting uniformly distributed to give the children a bright learning environment. The spaces of each floor integrate by the penetrating of light. It needs to shield and separate with light door barriers to achieve independent space partitions when meet different activities.

项目以向上生长的大树作为设计意象，并与树下学习的情景相呼应。这不仅体现了业主对子女成长的憧憬，而且体现了业主对子女教育用心良苦和无比重视。因此，在体现艺术性之余，大树拥有展现“爱的教育”的特殊功能。

项目为长向街屋，纵向较深，而横向过窄，再减去楼梯及房间，空间显得狭窄。设计师有意保留原建筑的挑空处，利用钢化玻璃配合树枝意象，巧妙创造出轻盈通透的楼梯扶手。这不仅能够营造出空间的宽敞感，而且使鲜明的绿色成为视觉焦点。另外，挑空处能够引入充足光线，解决多楼层所面临的光线不足问题。设计师更别具匠心地创造出随自然变化而变化的空中光井。

设计师通过减少各层隔间墙的手法降低光线与视觉上的障碍，仅保留最少量的墙面界定空间分区，以此维持空间的开放、穿透与连续的整体感。与此同时，将隔间墙上端切除，安装清透玻璃，可以达到室内各处都能看见大树的效果，同时又使光线均匀分布，给予小朋友明亮的学习环境。凭借光线的穿透性，各楼层空间实现相融。若遇到不同活动需要各自遮蔽，通过推拉轻薄门片，即可达到空间划分的效果。

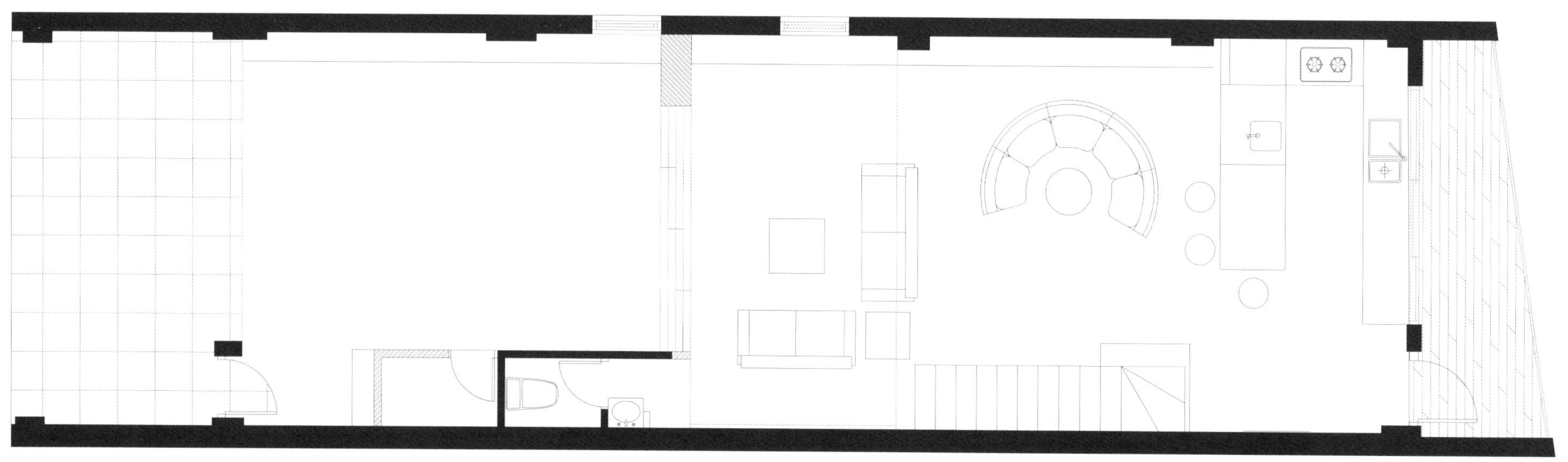

First Floor Plan
一层平面图

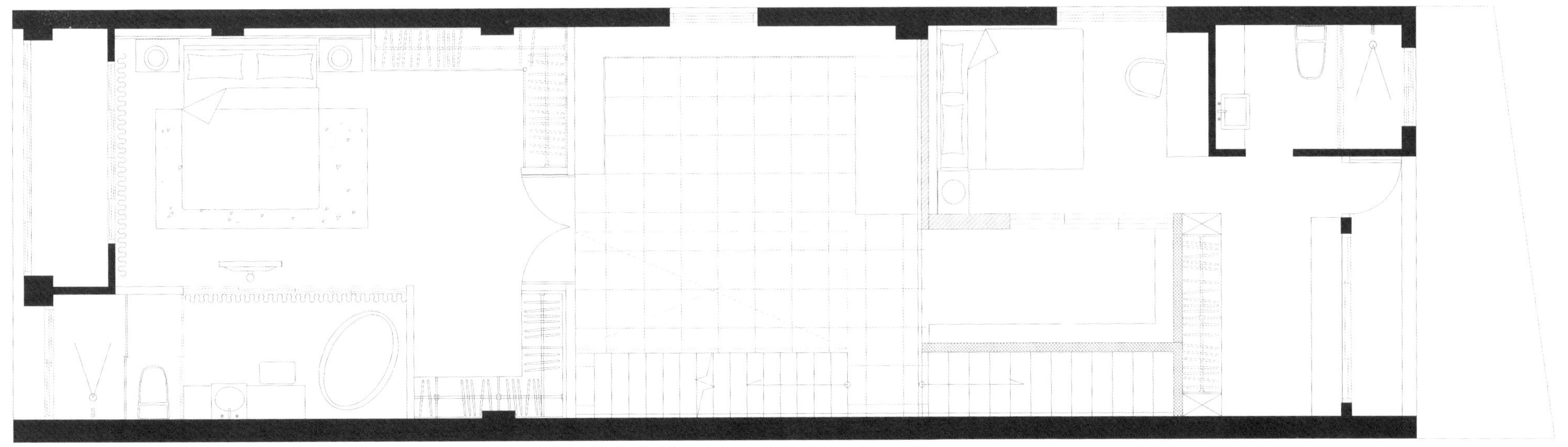

Second Floor Plan
二层平面图

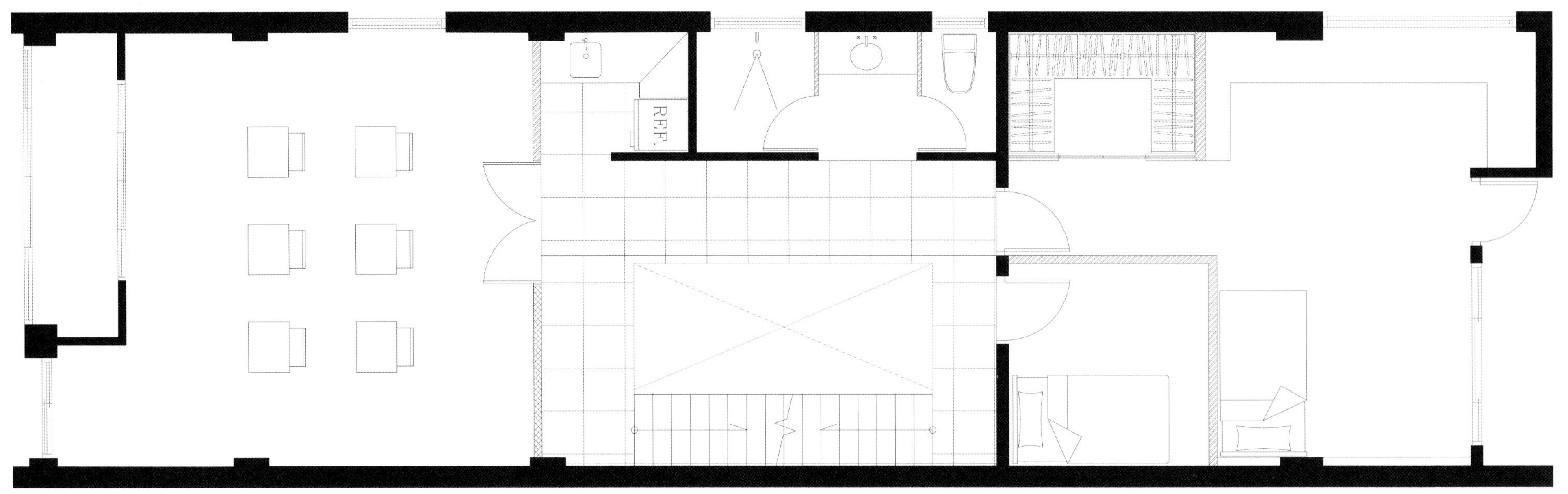

Third Floor Plan 三层平面图

Windsor Castle Phase III, Dongguan

东莞绿茵温莎堡三期

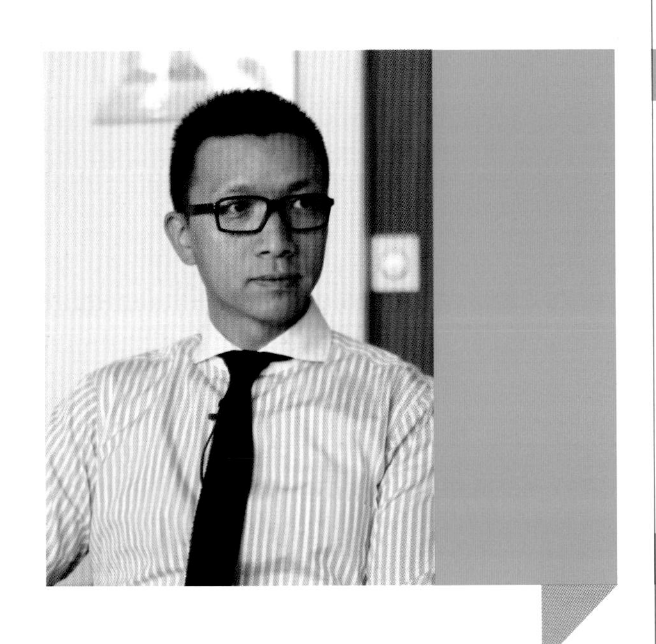

Designer: Rocky Chan, John Xit
Interior Design: Rocky Design
Location: Dongguan, Guangdong, China
Area: 603 m²

设计师：陈飞杰，薛永丰
室内设计：陈飞杰香港设计事务所
项目地点：中国广东省东莞市
面积：603 m²

Keywords 关键词

Romantic Theme 浪漫情怀

Classical Aesthetic 古典审美

Elegant Style 高雅气韵

Furnishings/Materials 软装 / 材料

Marble, Timber Veneer, Champagne Gold, Wired Glass, Solid Wood Floor, Hard Package, Soft Package, Art Wallpaper

云石、木饰面、香槟金、夹丝玻璃、实木地板、硬包、软包、艺术墙纸

Sometimes, space reflects the longing for a utopia, which serves as the home of one's sentiment, the extension of the thought, or the habitat for the soul. A pleasant space will relieve people of a whole day's fatigue, just like a melody in the morning which is elegant and pleasing.

As a single-family villa with a large courtyard, the project faces the challenge to well match the interior decoration with the outdoor landscapes. When planning the interior spaces, the designer takes advantage of the French window, glass door and patio to bring more daylight into the house. At the same time, the patio space is lifted, expanded and decorated by the stones and art works, to shape the contrast between light and shadow, possession and disclosure, and create an elegant, dignified and unique artistic space. The designer skillfully uses elegant brown, high quality leather, decorative bright surface materials and contrasting colors to create a textural and sophisticated living space. All the details are designed exquisitely to tell the stories of the villa and combine urban romance with modern living requirements, creating a kind of retro, fashionable, sophisticated and elegant lifestyle.

空间有时就是一种对乌托邦的寄想，可以作为情感的归宿和思想的延伸，或者是对心灵的收纳。一个赏心悦目的空间能驱散身体的疲惫，犹如清晨的乐章，高贵并且愉悦。

作为一个带有大面积庭院的独立别墅，如何将景观与室内设计做得相辅相成是项目的设计重点。在空间规划时，充分地利用了落地玻璃门窗，天井将光线引入室内。同时巧妙地运用中庭挑高扩宽空间维度并采用石材与艺术品的铺垫呼应，将古典审美范畴中的明暗对比，藏与露的比例予以现代的手法来演绎，充分营造出高雅、尊贵的气韵之余还融入不凡的艺术品格。

在对室内空间氛围的整体把握上，设计师匠心独运让雍容的咖啡色调弥漫整个空间，优质的皮面料，亮面装饰的点缀及局部对比色的运用丰富了空间，以高品质的饰面，精致的线条渲染着城市的轨迹，极具匠心的细致雕琢，婉约地诉说着每一个指尖碰触过的唯美故事，将都市的浪漫情怀与现代人对生活的需求相结合，营造出复古、前卫、精致的高雅生活。

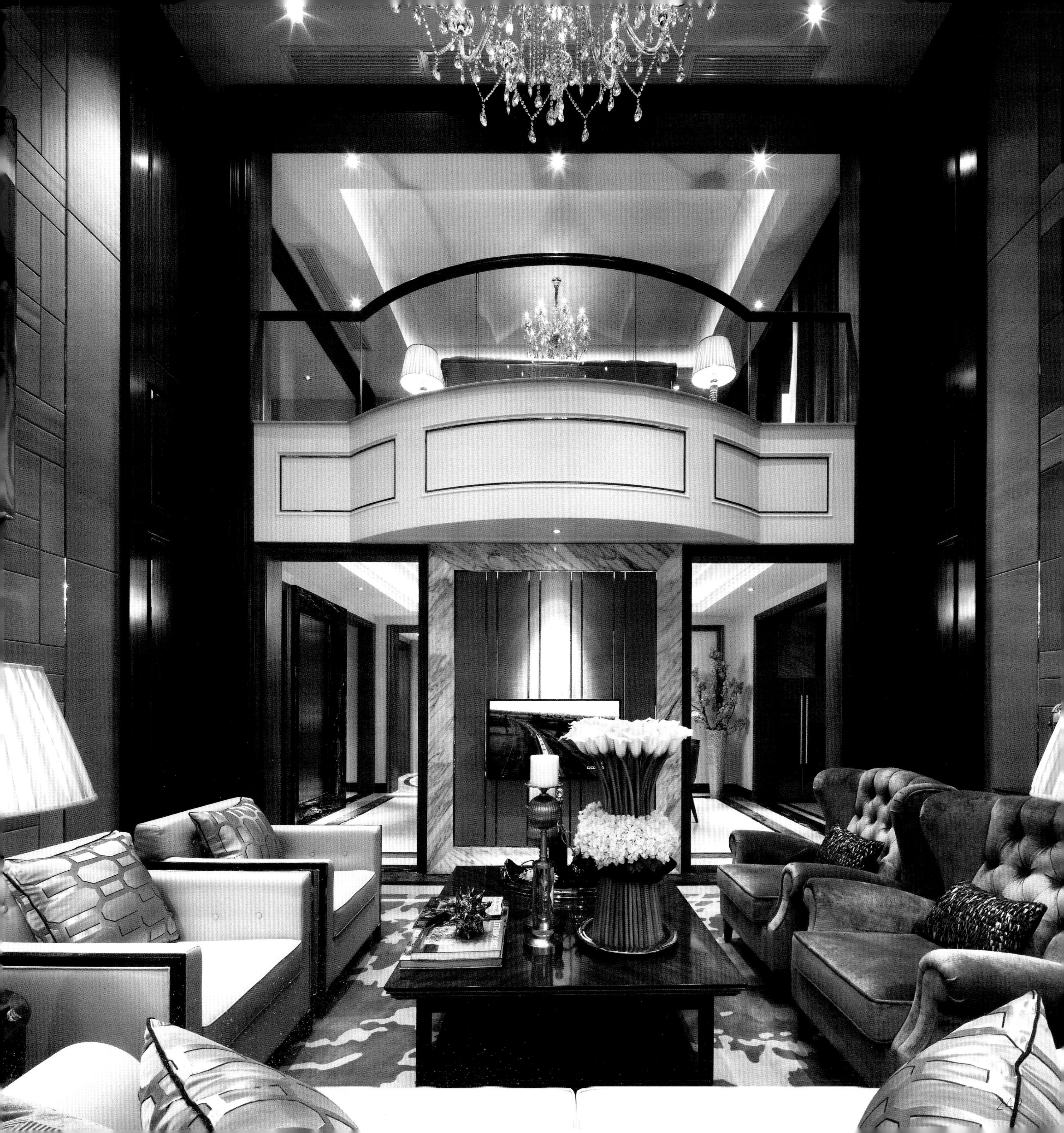

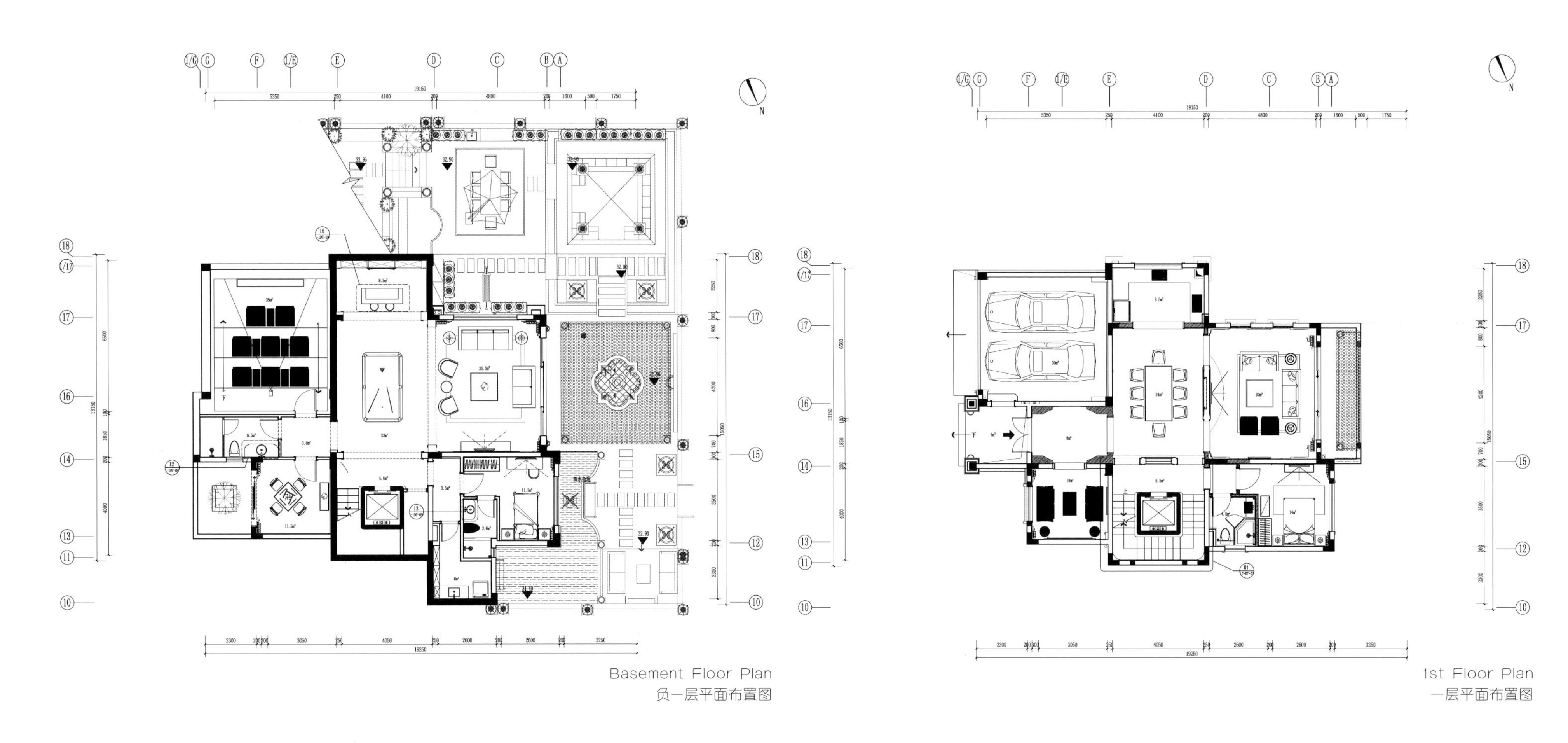

Basement Floor Plan
负一层平面布置图

1st Floor Plan
一层平面布置图

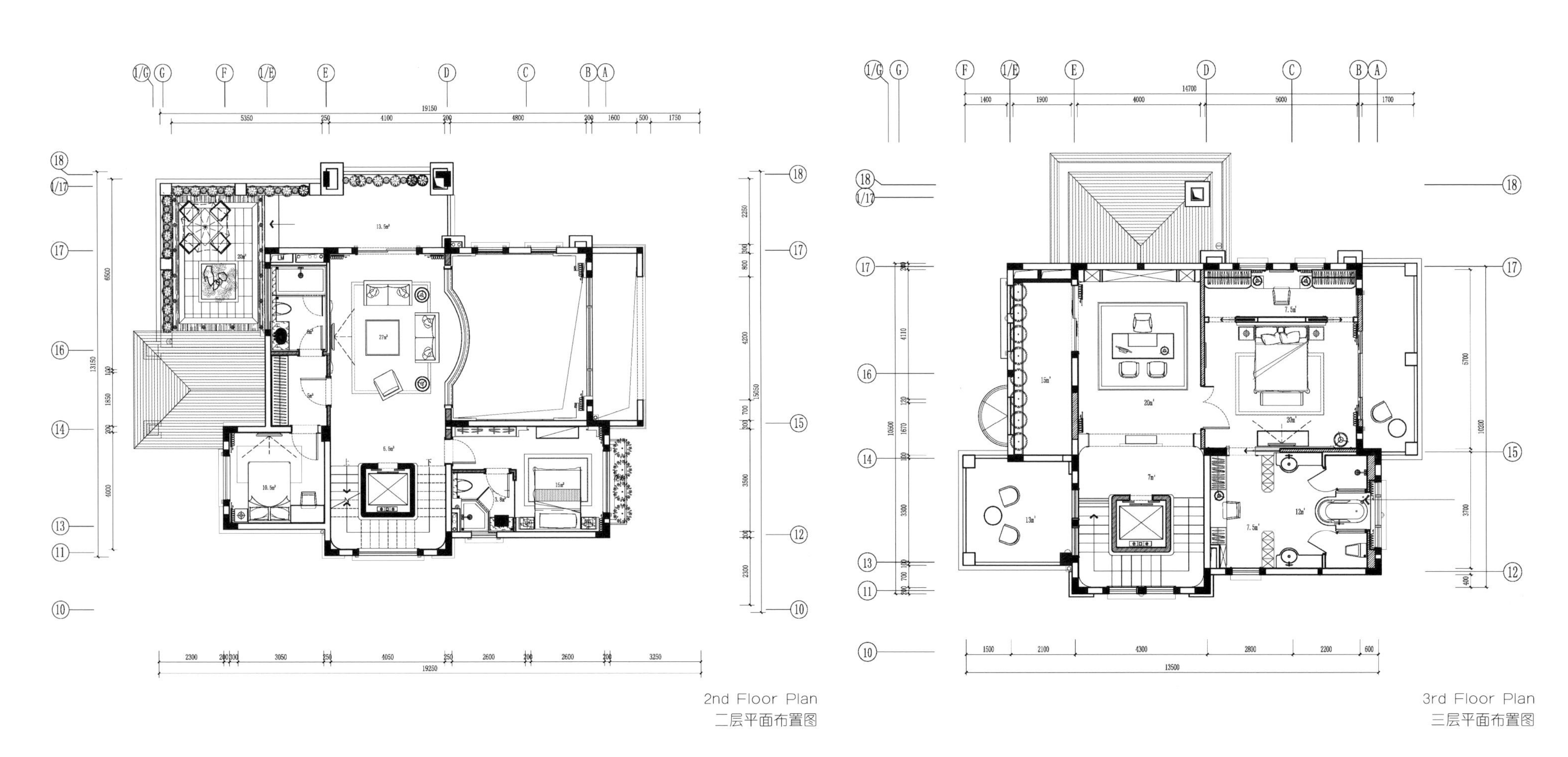

2nd Floor Plan
二层平面布置图

3rd Floor Plan
三层平面布置图

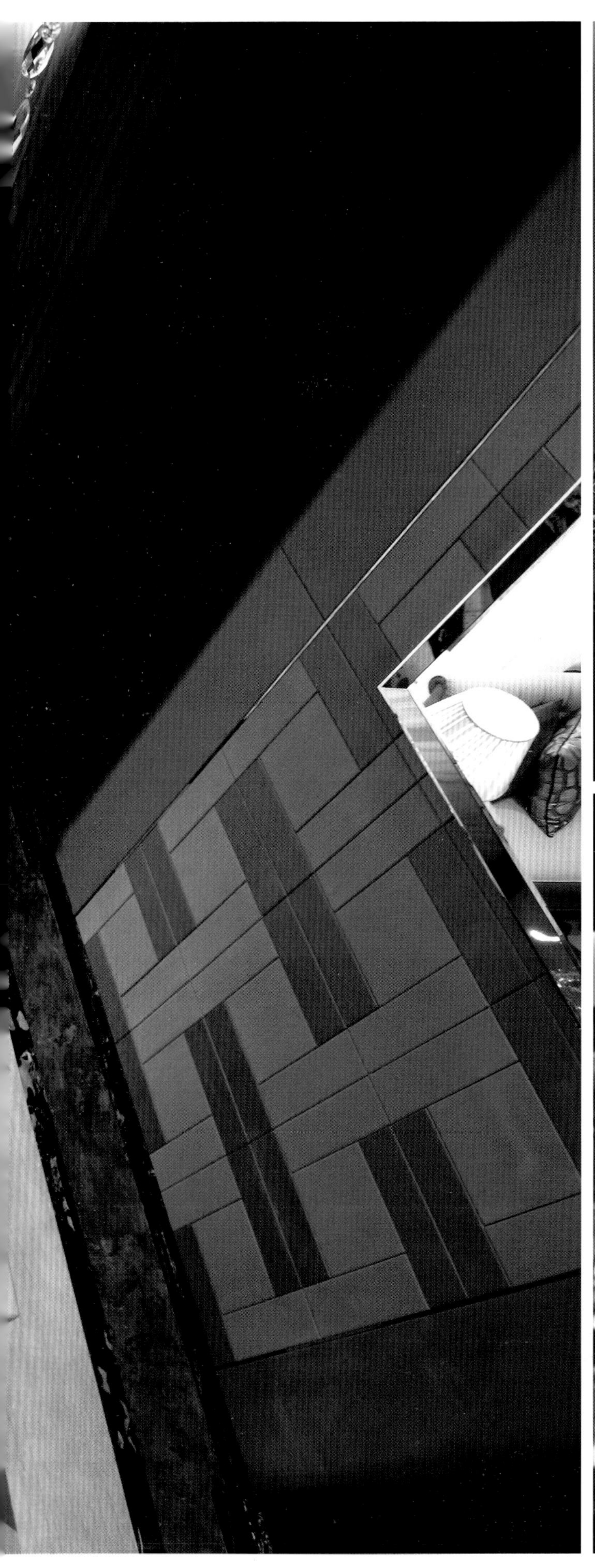

英国文化的七个时代(三)
亨利现在可以无所顾忌地

Metropolitan · Guiyang City Harbor

贵阳 · 中航城港湾

Designer: Simon Chong, Amy Du
Design Company: Simon Chong Design Consultants Limited
Location: Guiyang, Guizhou, China
Area: 346 m^2

设计师：郑树芬、杜恒
室内设计：SCD香港郑树芬设计事务所
项目地点：中国贵州省贵阳市
面积：346 m^2

Keywords 关键词

Quiet and Silent 清新安静

Unique Dynamism 非凡活力

Strong Stereo Sense 立体感强

Furnishings/Materials 软装 / 材料

Iron Art Trees, Mirrors, Crystal Lamps, Leisure Sofa, Plants, Art Paintings, Stripe Wooden Veneer

铁艺树、镜子、水晶灯、休闲沙发、绿植、艺术挂画、条纹木饰面

The totally four floored projects includes a basement garage with two cars, and family area, tea-tasting area, Western food kitchen, maid's room and so on, the first floor includes living room, dining room and recreation area, the second floor consists of study room and guest bedrooms, while master bedroom, terrace and outdoor SPA is placed on the third floor.

Stairs of each floor are fabulous. Because of high storey height, designer decorates the stairs with mirrors to avoid any sense of boring of which have made them fashionable and unusual. While the stairs of basement are decorated with two iron art trees which make master feel shocked. These bold but also creative methods adopted for the project have received the master's approval.

The most attractive decoration of the living room in the first floor is a large-sized wall oil painting. Moreover, the delicate crystal lamps, simple tea table, leisure sofa cooperated with a few plants have formed a beautiful landscape painting. The sofa is placed in an irregular way without making people feel strange but can enjoy the special feeling that the space provided.

One wall of the dining room is designed with wooden veneer, one wall is wallpaper, while the other two walls are decorated with art paintings and mirrors respectively to form the sense of contrast, but also bring a strong sense of modern. The nearby kitchen door adopts glass sliding door which becomes a unity when the door is closed.

Bringing a sense of fresh and quiet, grey pattern wallpaper and comfortable furniture combination in the study brighten the space. All kinds of furnishings collected in wooden bookcase show extraordinary vitality.

The master bedroom featured with striped wood veneer decorated as background wall, and the two black-and-white ballet frames show the extreme elegance. Beige beddings and the tone of the room contrast with each other. Light green stripe art carpets create unique space.

Large amount of middle colors and wooden cabinets in guest bedrooms create soft and comfort feeling of the space. While the modern abstract paintings and green striped carpets seem harmonious and also enrich the whole space.

Since children' world is always like as vibrant as things to grow, designers mainly use grass green tone for it. The beddings and bolsters are in dot patterns which are with strong stereo sense but also with the characteristics of spaciousness and abundance.

项目共有四层，负一层为两个车库，从车库走进来便是家庭厅、品茶区、西橱、工人房等；一层为客厅、餐厅、休闲区；二层为书房、客房；而三楼便是主人房及天台休闲区及露天SPA。

值得一提的是每一层楼梯的设计。由于层高较高，为避免单调，设计师在每一层的楼梯间都用镜子做装饰，时尚而独特。负一层楼梯处采用两棵铁艺树作装饰，创造了一种震撼的效果。这样大胆而创新的手法获得到了户主的高度认可。

一层客厅最吸引眼球的莫过于一幅巨大的墙面油画。另外，精致的水晶灯、简单的茶几与休闲沙发、几棵绿植构成了一幅美丽的风景画。沙发以不常规的方式摆放，却不会使人感到突兀，反而使人更能享受空间的特别之处。

餐厅中一面是木饰墙面，一面是墙纸，两面以艺术挂画和镜子营造对立感，并且使餐厅的现代感更加强烈。紧邻的厨房门采用了夹丝玻璃推拉门，拉上门时完全成一体，完全没有违和感。

书房带来的是一股清新安静之感。灰色花纹壁纸和舒服的家具组合使空间变得明亮。木质书柜收藏的各类陈设品显出了非凡的活力。

主卧非常有特色。背景墙以条纹木饰面为装饰，而两幅黑白芭蕾舞蹈画框将优雅风情推到了极致。米黄色的床品与房间色调相互映衬。浅绿色条纹艺术地毯让整个空间别具一格。

客卧运用大量的中色系和木质柜营造出柔和安逸的空间感。而现代抽象画和绿色条纹的地毯非但不显得突兀，还使这个空间饱满、丰富起来。

儿童的世界里总是像万物生长般那样充满生机，设计师在这里运用了以草绿色为主调。床品及抱枕以小波点凸显，不仅立体感强，而且活泼可爱。

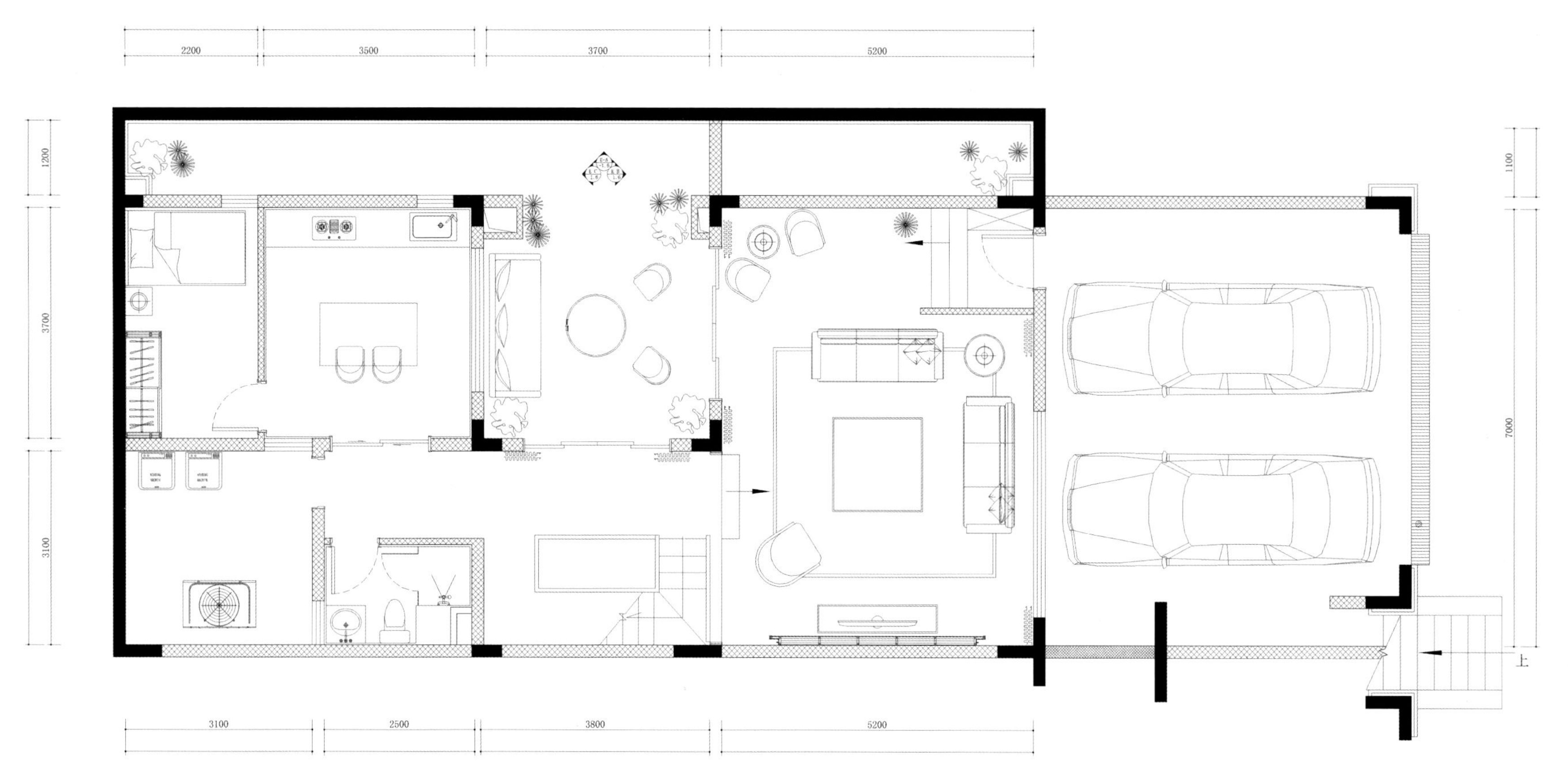

Basement Floor Plan 负一层平面图

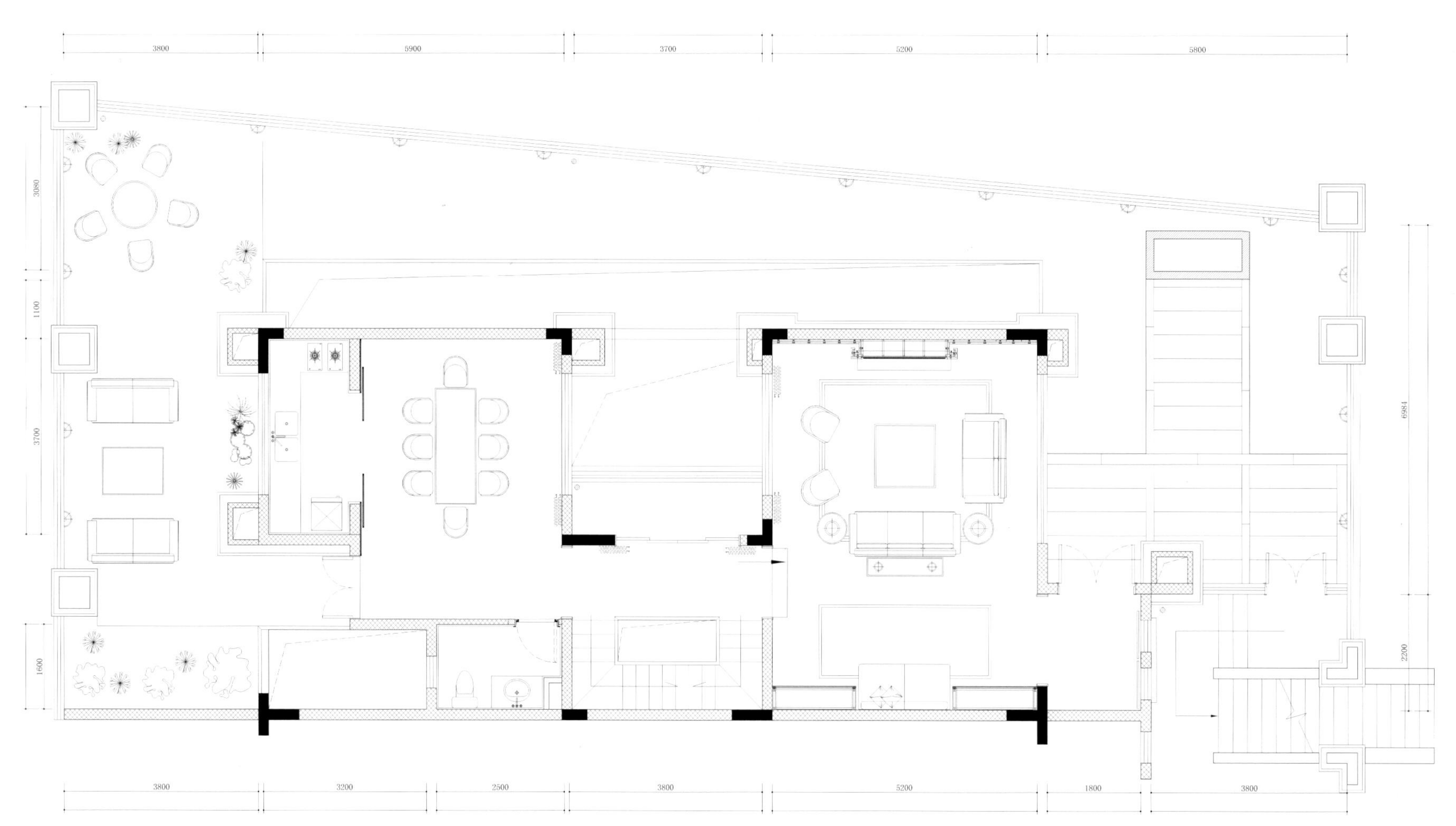

First Floor Plan 一层平面图

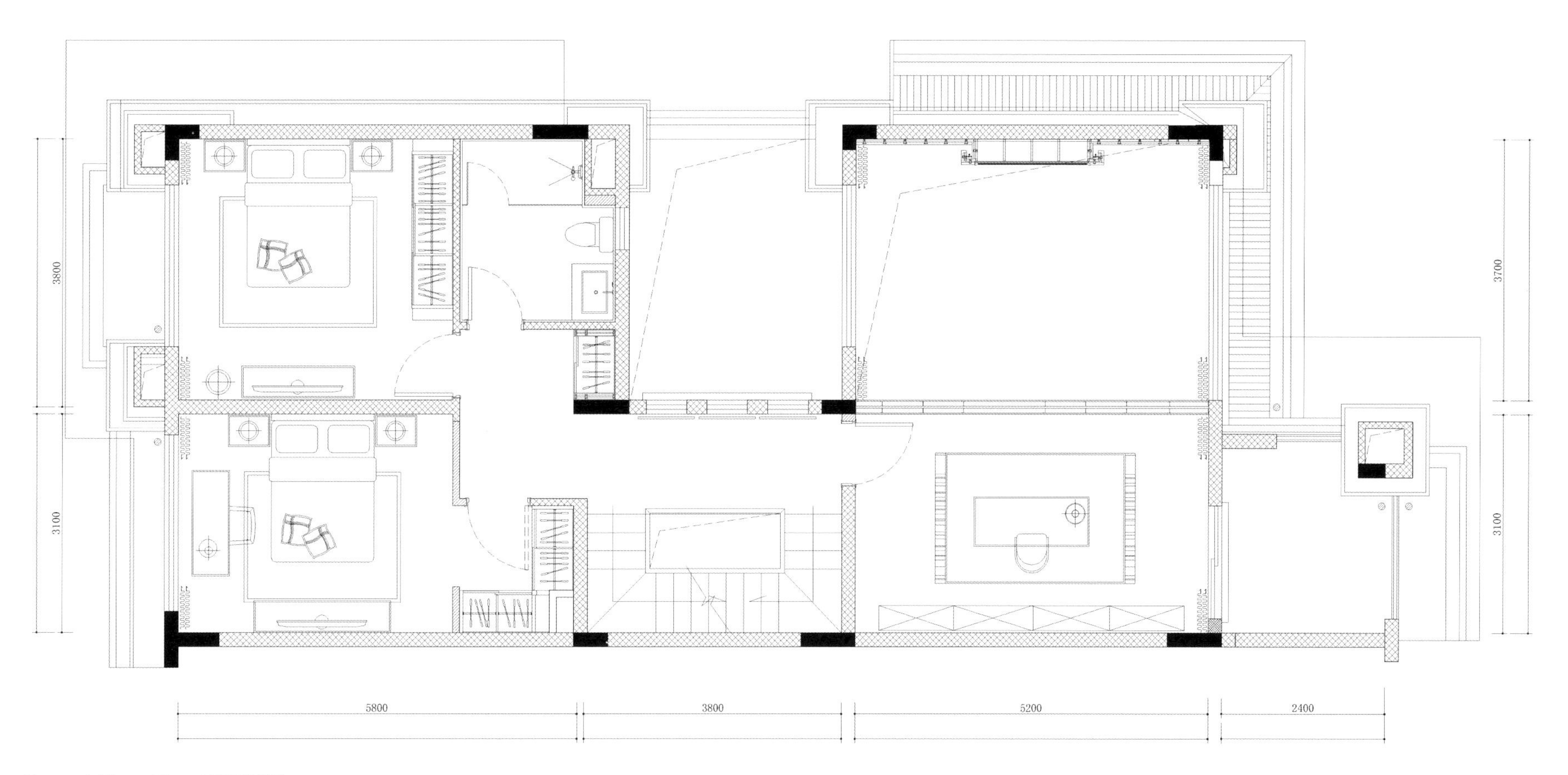

Second Floor Plan 二层平面图

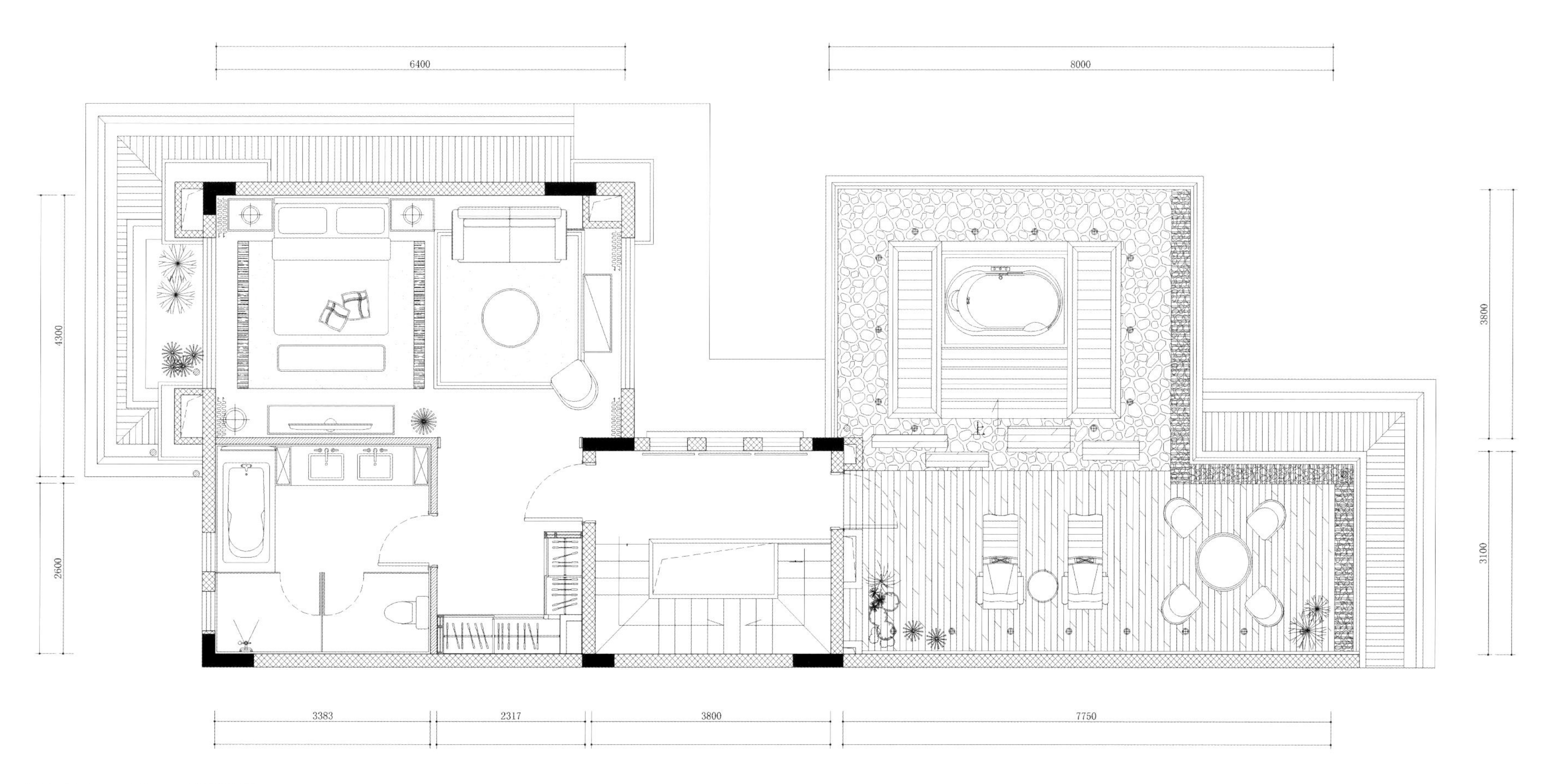

Third Floor Plan 三层平面图

SCD
KELLY HOPPEN HOME
SAKURA
LUXURY HOTELS

THE LOVELY BONES
Furniture Decoration

THE LOVELY BONES
AMANG

Other Styles

其他风格

Concise
质朴简约

Vigorous
生机盎然

Vibrant
富有活力

Changsha CSC Meixi Lake Era Villa Sample Room (Baroque Love Song)

长沙中建梅溪湖一号别墅样板间(巴洛克恋曲)

Designer: Liu Weijun
Interior Design: PINKI Creative Group & Liu and Associates (IARI) Interior Design Co., Ltd.
Location: Yuelu District, Changsha, Hunan, China
Area: 580 m^2

设计师：刘卫军
室内设计：PINKI品伊创意集团&美国IARI刘卫军设计师事务所
项目地点：中国湖南省长沙市岳麓区
面积：580 m^2

Keywords 关键词

Baroque Style 巴洛克风格

Luxury and Exquisite 奢华精致

Stereo Structure 立体结构

Furnishings/Materials 软装 / 材料

Volakas White Marble, New Moon Marble, Archaized Brick, Archaized Wood Floor

爵士白大理石、新月亮大理石、仿古砖、仿古木地板

The sample room takes Baroque Love Song as the theme, meaning the spirit movement blending the baroque classical art and modern life. This sample room is mainly for the successful man who is romantic and is longing for the western splendid culture, and the householder who has western life experiences and loves western classical art.

The design combines the luxury and noble elements in European baroque style and the detail elements in modern life, making this project become a smart and lively life form and making people experience this luxury and pleasant living environment. Designers believe that a successful design is close to human nature and can be blended into their lives. The life feelings are combined together like a piece of gorgeous aria.

From the life view, the design offers the best environment for the life in terms of stereo structure. This project is offered with the most distinctive function layout, from the underground entertainment room to the luxury audio-video room, from the wine cellar to the living room bar, the capacious swimming pool and the comfortable courtyard, which make people experience the one-stop high-end life at home. At the same time, the design links every space skillfully, following the life rhythm and being close to the life, making people experience the luxury and comfortable life.

The utilization of the technique of building stereo structure with detailed lines and the use of leather and metallic materials in design show a luxury, mighty and exquisite interior space. All those flash cambered classical elements, classical sculpture, noble and exquisite fabrics show the luxury and delicacy of European classical culture and modern life.

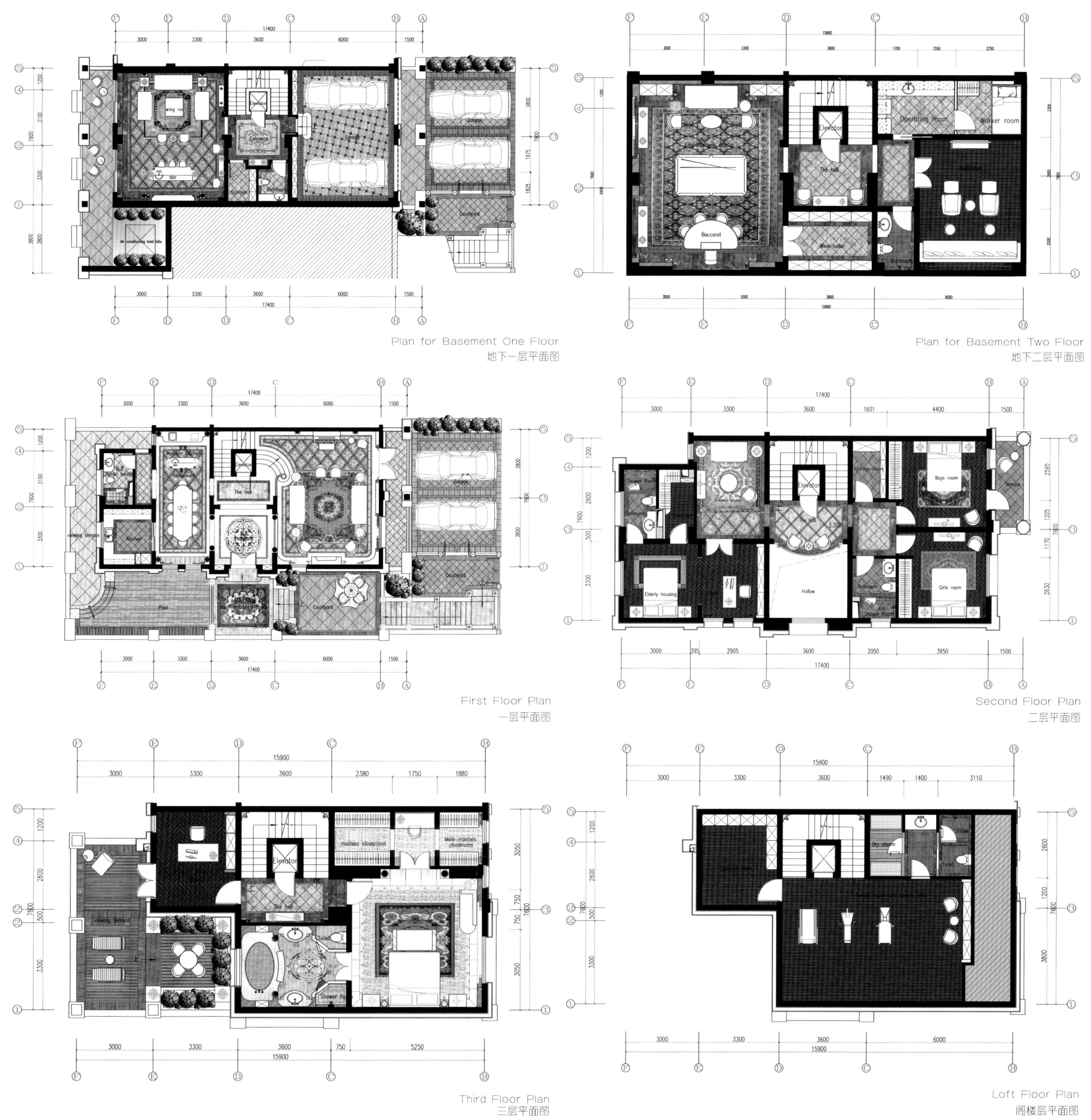

Plan for Basement One Floor
地下一层平面图

Plan for Basement Two Floor
地下二层平面图

First Floor Plan
一层平面图

Second Floor Plan
二层平面图

Third Floor Plan
三层平面图

Loft Floor Plan
阁楼层平面图

项目样板间的主题为巴洛克恋曲，意为经典的巴洛克古典艺术与现代生活交融的精神乐章。样板间主要针对有着浪漫情怀、向往西方灿烂文化的成功人士以及有着西方生活经历、热爱西方古典艺术的户主。

设计撷取欧洲文化中巴洛克风格的奢华、高贵，结合现代生活的细节元素，赋予项目灵动鲜活的生命形态，让人们在奢华中感受舒适宜人的居住环境。设计师坚信成功的设计是贴近人性、融入生活的。将生活赋予的情感融合到一起，宛若一曲华丽的咏叹调。

空间布局上，从生活的角度出发，立体化地为生活提供最优质的环境。从地下娱乐厅到豪华影音室，从酒窖到会客厅酒吧，宽敞的泳池、舒适的庭院，设计在功能上给予本案最具特色的布局，让人们在家中体验一站式高端生活。在功能丰富的同时，巧妙地联动每一个空间，跟随生活的节奏，贴近生活，感悟奢华与安逸。

设计选材中，运用细节线条铸造立体结构的手法，配合皮革和金属感材料的使用，展现出奢华、强势、精美绝伦的室内空间。在白色温柔里，那些带着闪光弧度的古典元素、经典雕刻及高贵且细腻的面料无不衬托着欧洲古典文明与现代生活的奢华与精致。

LIVING WINE

QUEEN
KING
BISHOP

Les Miserables
LES MISERABLES

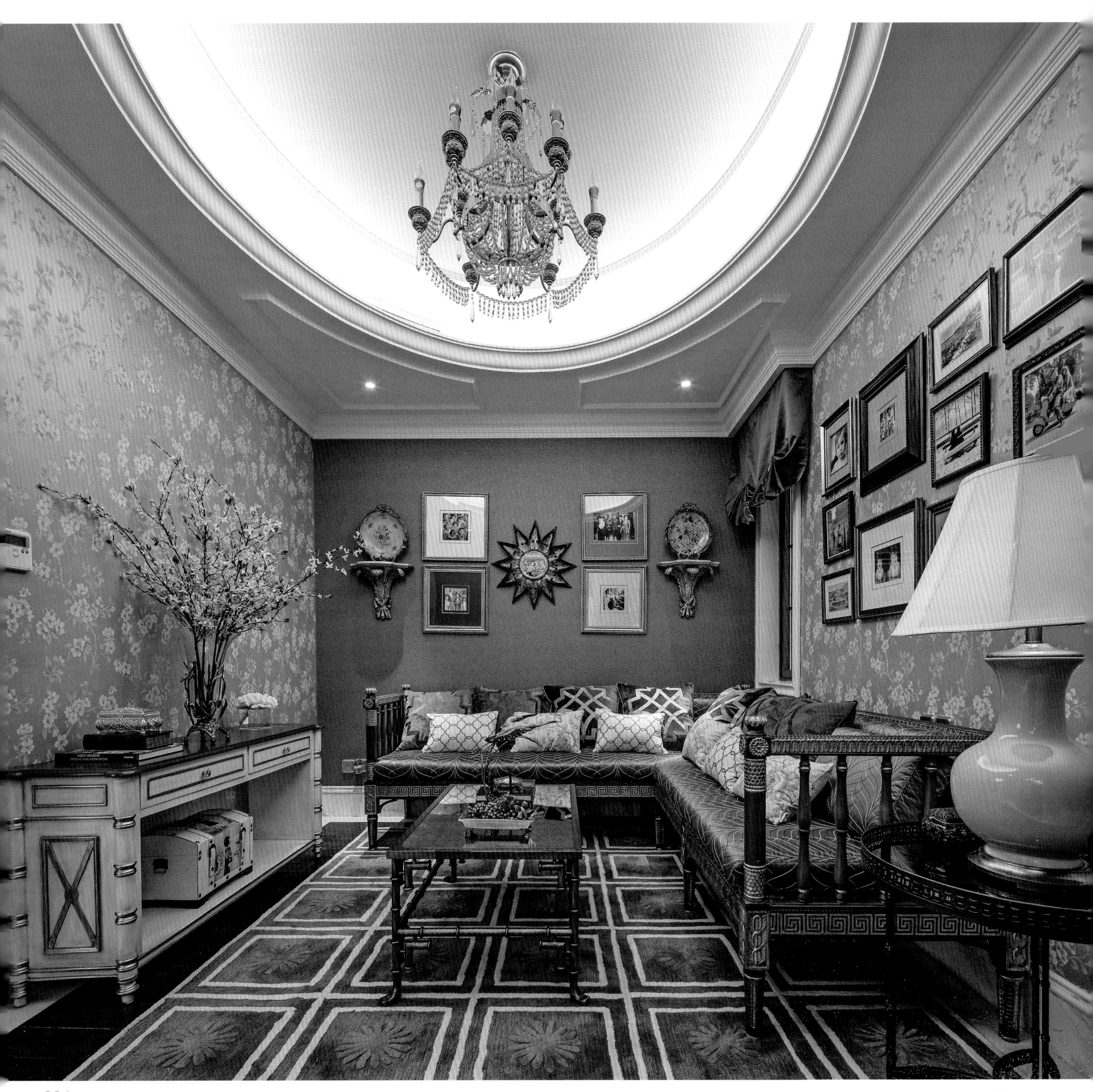

Poly Silver Beach Villa Zone Q

保利银滩 Q 区别墅

Designer: Tony Ho
Interior Design: Daosheng Design Co., Ltd.
Location: Hailing District, Yangjiang, Guangdong, China
Area: 253 m^2

设计师：何永明
室内设计：广州道胜设计有限公司
项目地点：中国广东省阳江市海陵区
面积：253 m^2

Keywords 关键词

Southeast Asian Style 东南亚风格

New Chinese Elements 新中式元素

Exotic Style 异域风情

Furnishings/Materials 软装 / 材料

Paris Grey Marble, Beige Travertine Marble, Rosy Gold Stainless Steel, Gold Foil

巴黎灰大理石、黄洞石大理石、玫瑰金不锈钢、金箔

Designed on the basic concept of hearing the ocean, the villas are set around by the blue ocean and sky. When entering into the unit, the views outside the French windows are catching the people's eyes, while they can enjoy the happiness of family. People, who are used to the busy and hurried urban lives, are quite eager for the relaxation of body and soul, while the simple Southeast Asian style will help them to forget the worries and sorrows for the moment, bringing them the happiness of closing to the nature and the ultimate eased living atmosphere. Spacious space with great streamline, the lines and the surfaces are well cooperated to achieve the harmony between the spaces and environment.

The concept of Southeast Asian design comes from the nature, which adopts wood and stone to create the natural bedrooms with characteristic of new Chinese style, and uses simple and concise designs instead of complicated moldings. The streamlines of the furniture are stretched and graceful, simple but smooth, just explained the elegant decorative style.

Exotic accessories have enriched the visual content and space impact, besides, they can also highlight the simplicity of the Southeast Asian style and the quietness of the Chinese style, making the spaces more romantic.

The designer selects a little rosy red for the flowers in this simple color tone, just like integrating lots of Buddhist allegory with the flowers, warm and bright. In addition, they have also added the exotic style and Buddhist elements to the unit, quiet but full of philosophy.

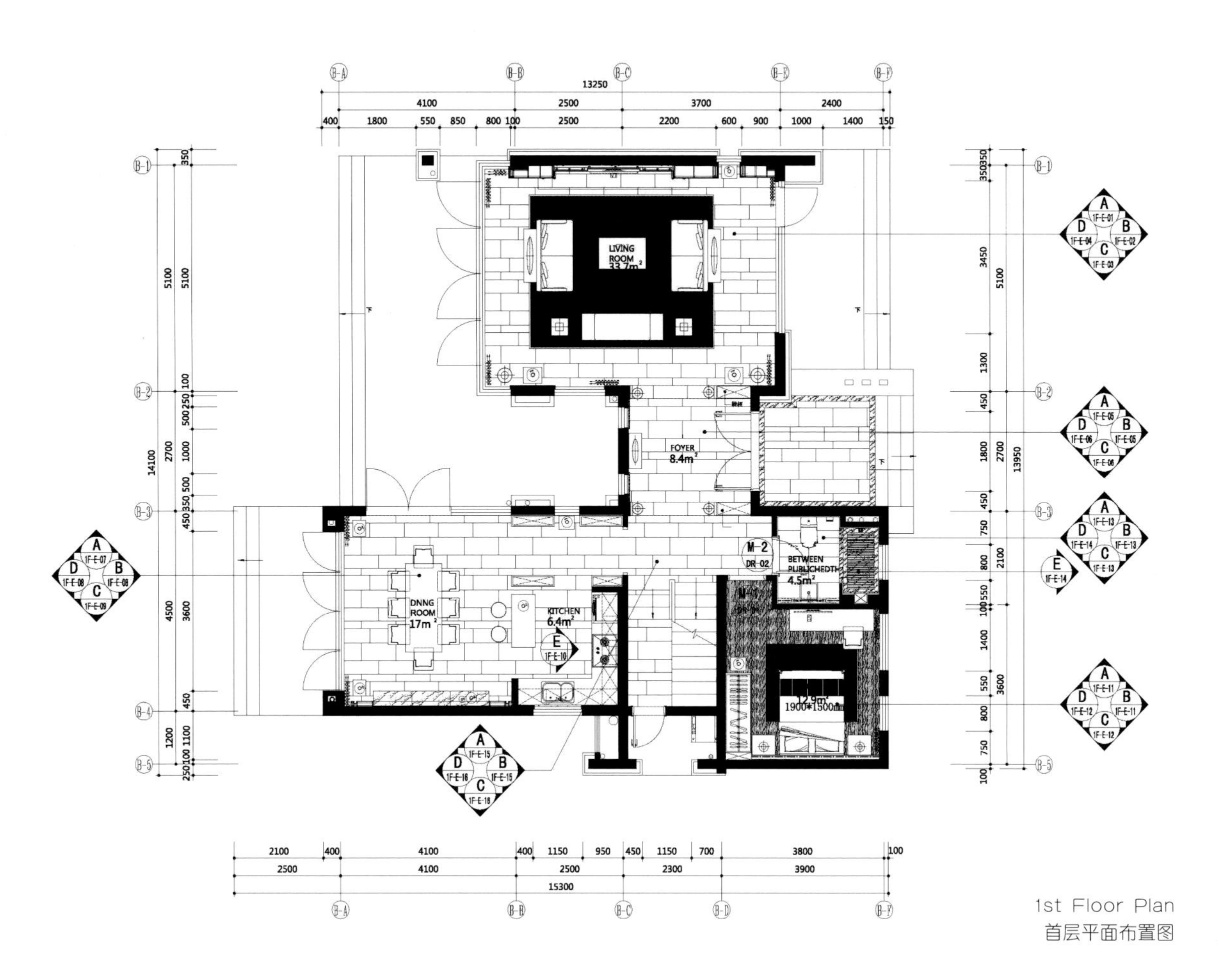

1st Floor Plan
首层平面布置图

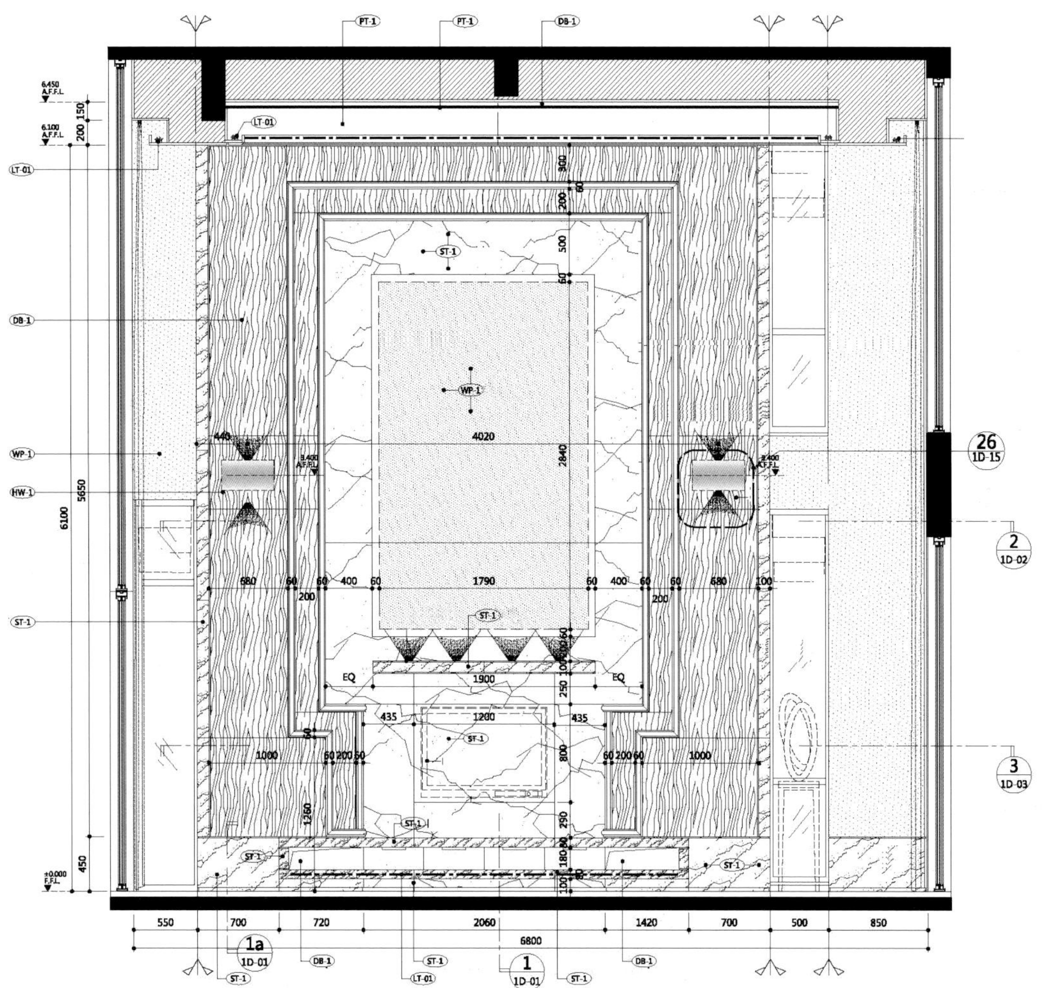
Elevation A of Sitting Room on Ground Floor 首层客厅A立面图

设计以听海为主题，坐拥碧海蓝天。进入室内，即被落地窗外的院景所吸引，一家人其乐融融，享受天伦之乐。久居都市匆忙而繁碌的人们都希望得到心灵的舒缓和灵魂的释放。朴实自然的东南亚风格让人暂时忘掉所有的烦恼和忧伤，带来亲近自然的快感和绝对放松的居家气氛。平面开敞流动，空间用线以及由线构成的面连贯穿插，注重空间的交互性和环境的相互融合。

东南亚风格的设计取材自然、独树一帜，用木材与石材来打造居室的自然之美和略带新中式的民族特色，抛开繁复的装饰线条，用简单整洁的设计取而代之。家具线条舒展而优美、简洁而流畅，犹如行云流水，演绎了庄重而优雅的风格。

富有异域风情的饰品，既丰富了空间的视觉内容，又增加了空间的影响力，突显出东南亚风的质朴以及新中式的静谧，将空间打扮得极富情调。

在质朴和谐的色调中，在花艺上用少量艳丽的玫红色，仿佛藏着无数的禅意，使人感到温馨、愉悦，别有一番风情，使空间散发着淡淡的异域气息，同时也让空间禅味十足，静谧而蕴藏哲理。

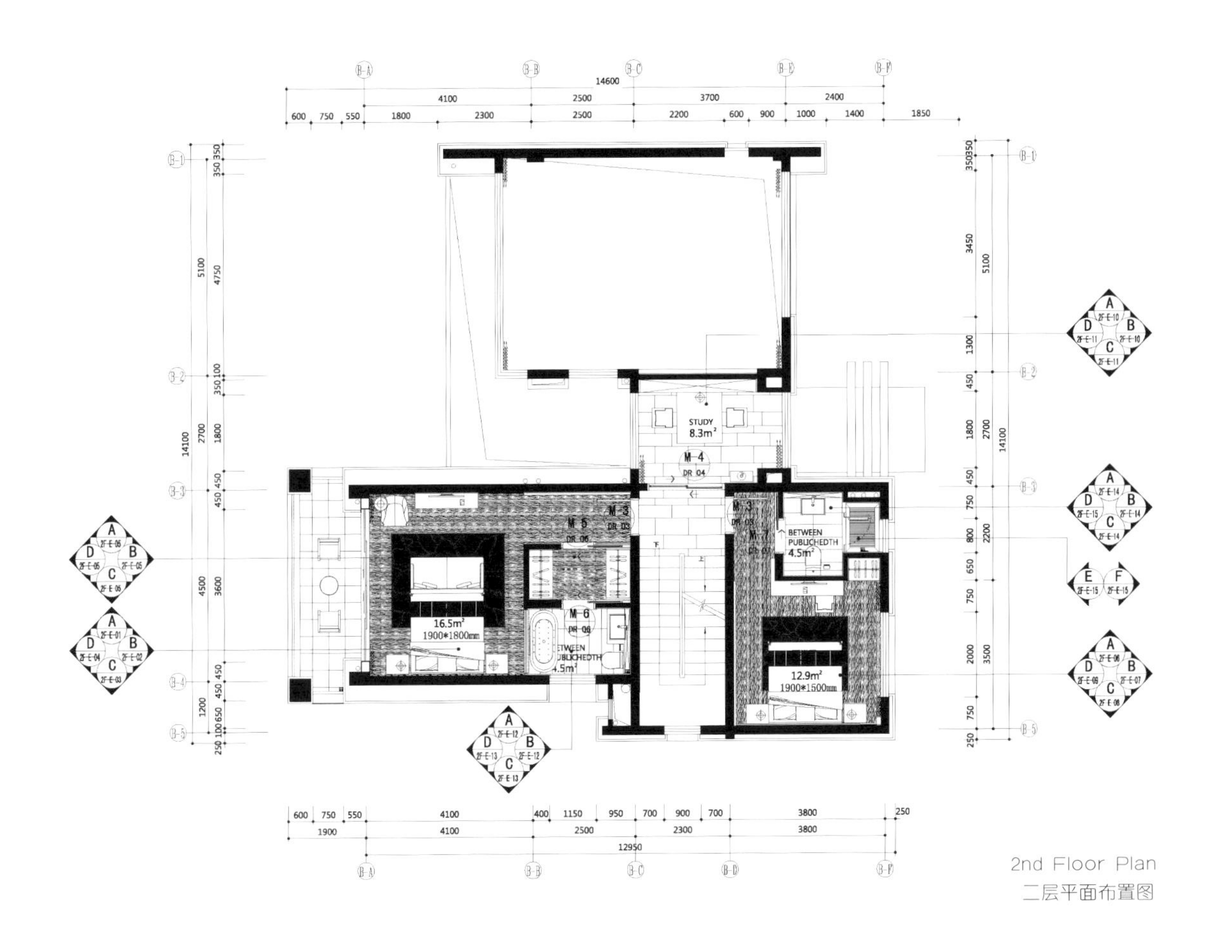

2nd Floor Plan
二层平面布置图

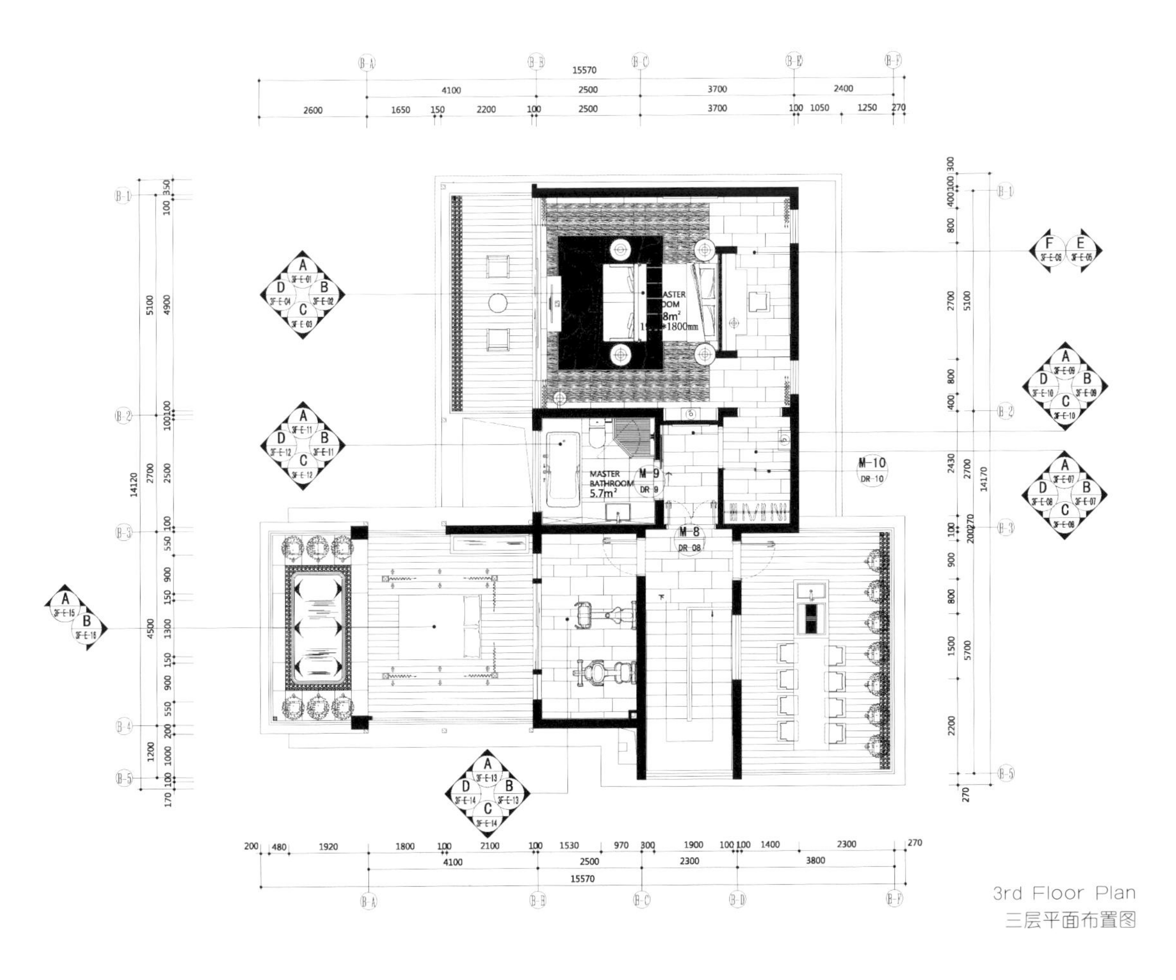

3rd Floor Plan
三层平面布置图

Shenyang Zhonghai City (Sensational Simiane Style)

沈阳中海城（迷情施米雅）

Designer: Danfu Liu
Interior Design: PINKI Creative Group & Liu and Associates (IARI) Interior Design Co.,Ltd
Location: Shenyang, Liaoning, China
Area: 437 m^2

设计师：刘卫军
室内设计：PINKI品伊创意集团&美国IARI刘卫军设计师事务所
项目地点：中国辽宁省沈阳市
面积：437 m^2

Keywords 关键词

Castle Style 古堡风格

Mediterranean Element 地中海元素

Pastoral Style 田园风情

Furnishings/Materials 软装 / 材料

Paint, Stone, Wallpaper, Leather, Hill-grain Oak Veneer Embellishment, Antique Teak wooden Floor, Tile

涂料、石材、墙纸、皮革、橡木山纹饰面板索色、柚木仿古木地板、瓷砖

The project is in the castle style of Simiane-la-Rotonde with simple and relaxing lifestyle integrated in the interior design. There are four layers and a basement with a distinctive leisure area. A wild lion's head is decorated on the wall of natural stone, and gurgling water pours from the lion's mouth into the swimming pool made of mosaic and marble. The various plants on both sides of the pool form a luxurious sense of the royal nobles.

The huge pastoral painting on the wall of the ground floor's living room is far from conventional and fully embodies the lavender life of Provence. The mimetic stove on the other side of the wall sketches out the European classical breath. The deer fresco above the stove is quite interesting and throughout the leisure style of the whole project. Meanwhile, the design of void structure is full of spaciousness. The arched Roman column and the door opening design makes the living room distinctive. Iron art decoration, crystal chandelier and heavy solid wood furniture provide the whole space with heavy elegance and grace. And the use of Mediterranean elements adds certain vigor and vitality.

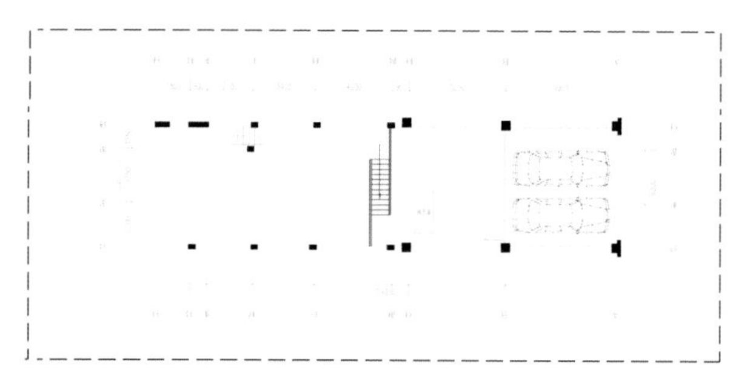

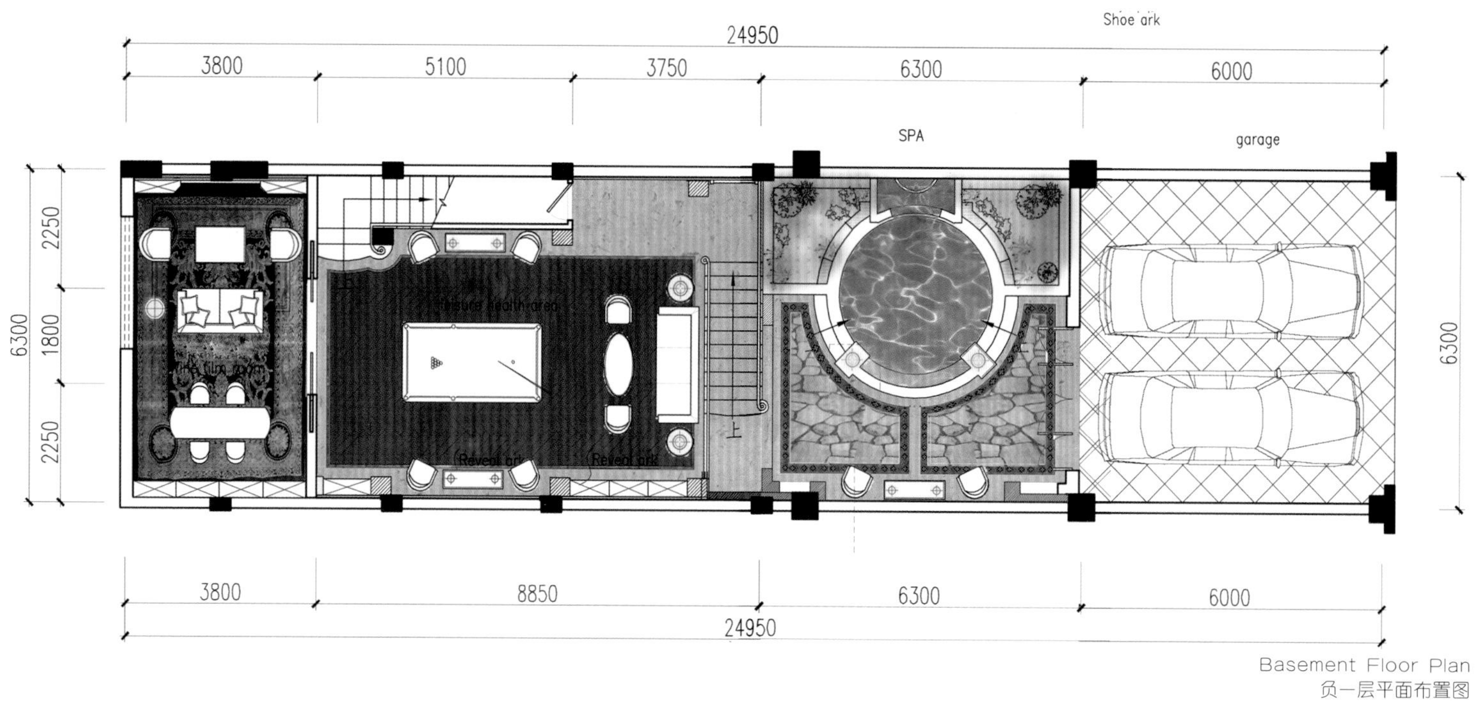

Basement Floor Plan
负一层平面布置图

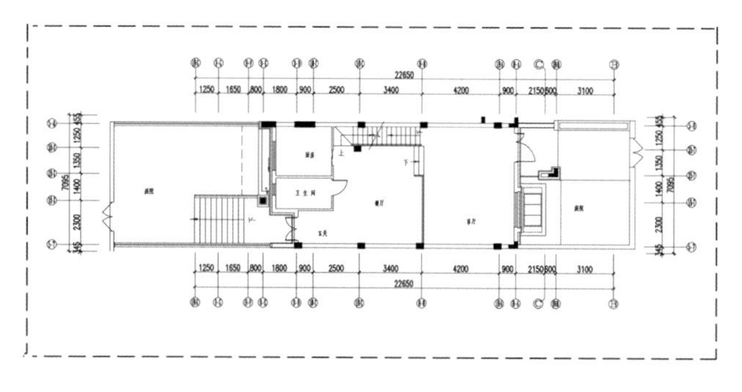

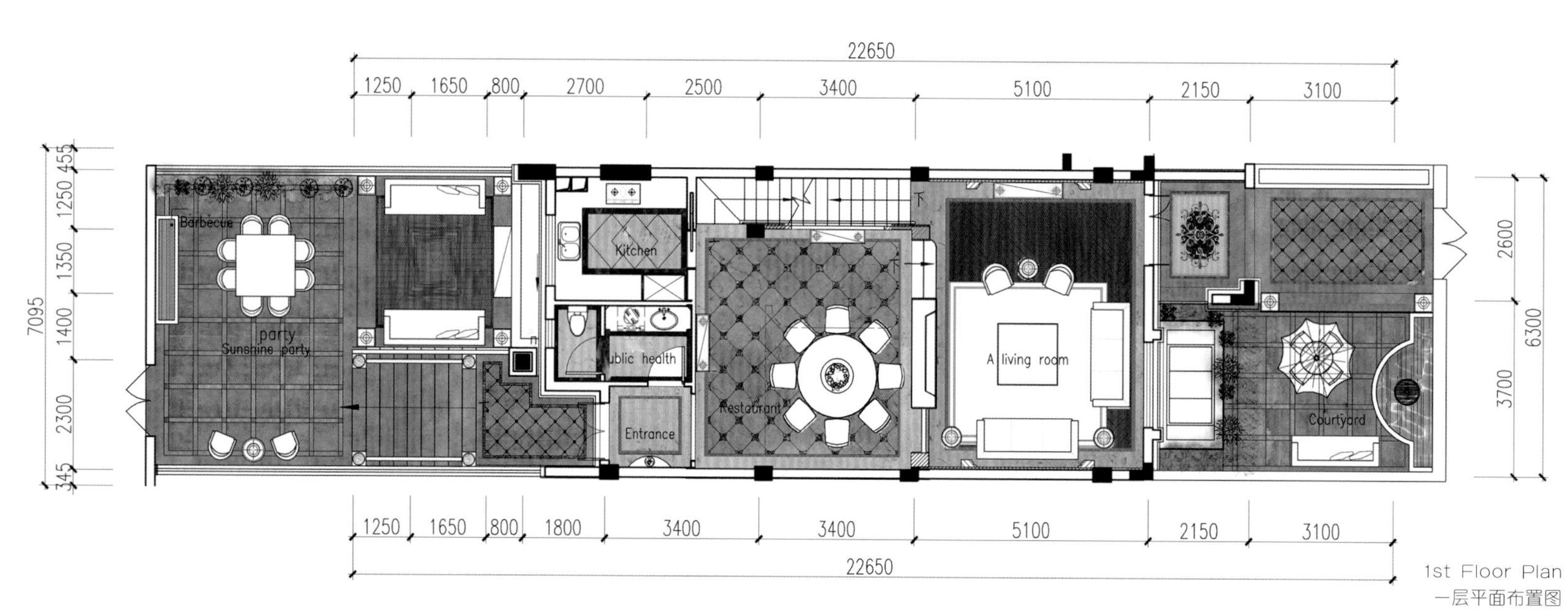

1st Floor Plan
一层平面布置图

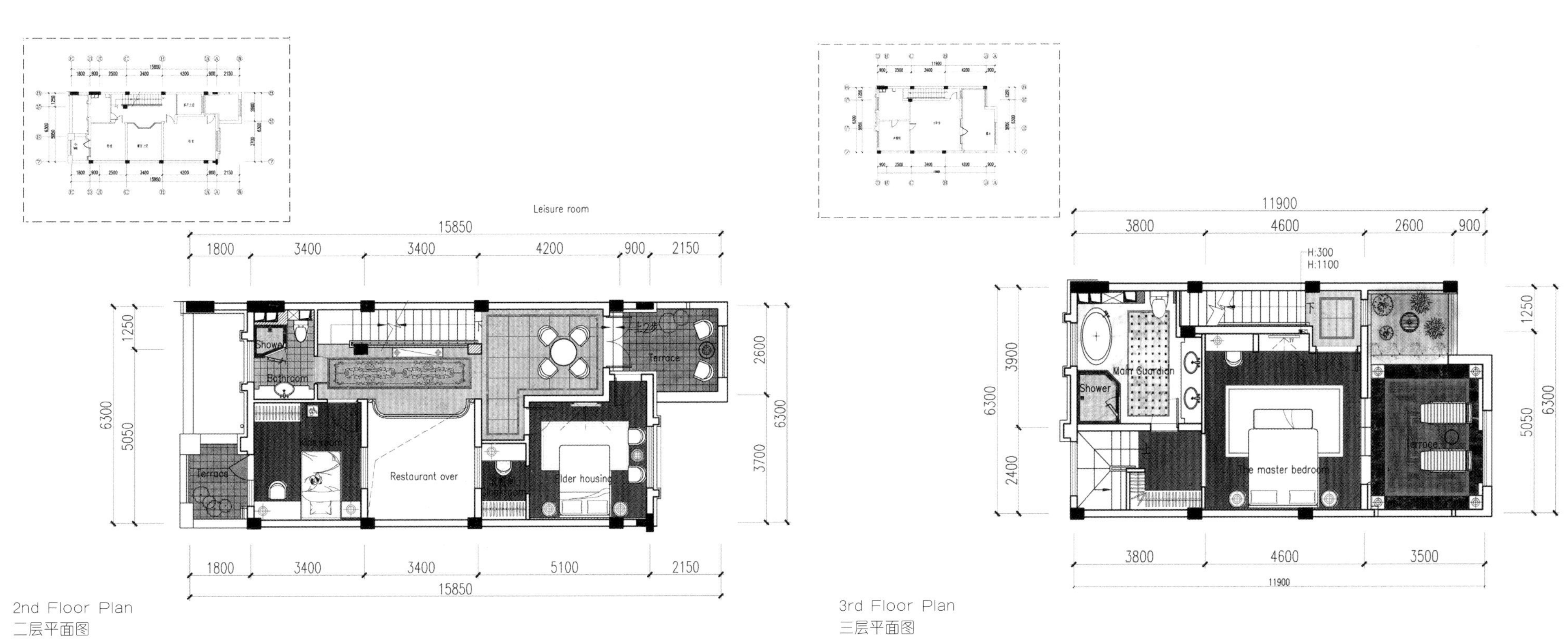

2nd Floor Plan
二层平面图

3rd Floor Plan
三层平面图

项目遵循施米雅小镇的古堡风格进行设计。设计师将简单无忧、轻松慵懒的生活方式融入室内设计当中。项目共设有四层，外加一个地下层。

负一层的悠闲区极具特色。原石堆砌的墙面上装饰着一个狂野的狮头。潺潺流水从狮口中涌出，缓缓注入由马赛克和大理石砌成的泳池中。另外泳池两旁栽种着各色植物，俨然营造出宫廷贵族的奢华感。

首层客厅以巨大的田园油画作为背景墙，既不落俗套，又将普罗旺斯的薰衣草生活表现得淋漓尽致。而另一面墙仿制出火炉，勾勒出欧洲的古典气息。火炉上方的野鹿壁画更别有一番趣味，贯穿整个项目的闲适风格。同时，挑空的设计让客厅空间感十足。拱形的罗马柱和门洞设计令客厅里别有洞天。铁艺装饰、水晶吊灯和厚重的实木家具赋予整个空间沉稳、优雅和大气。地中海元素的使用为其增添几分生机和活力。

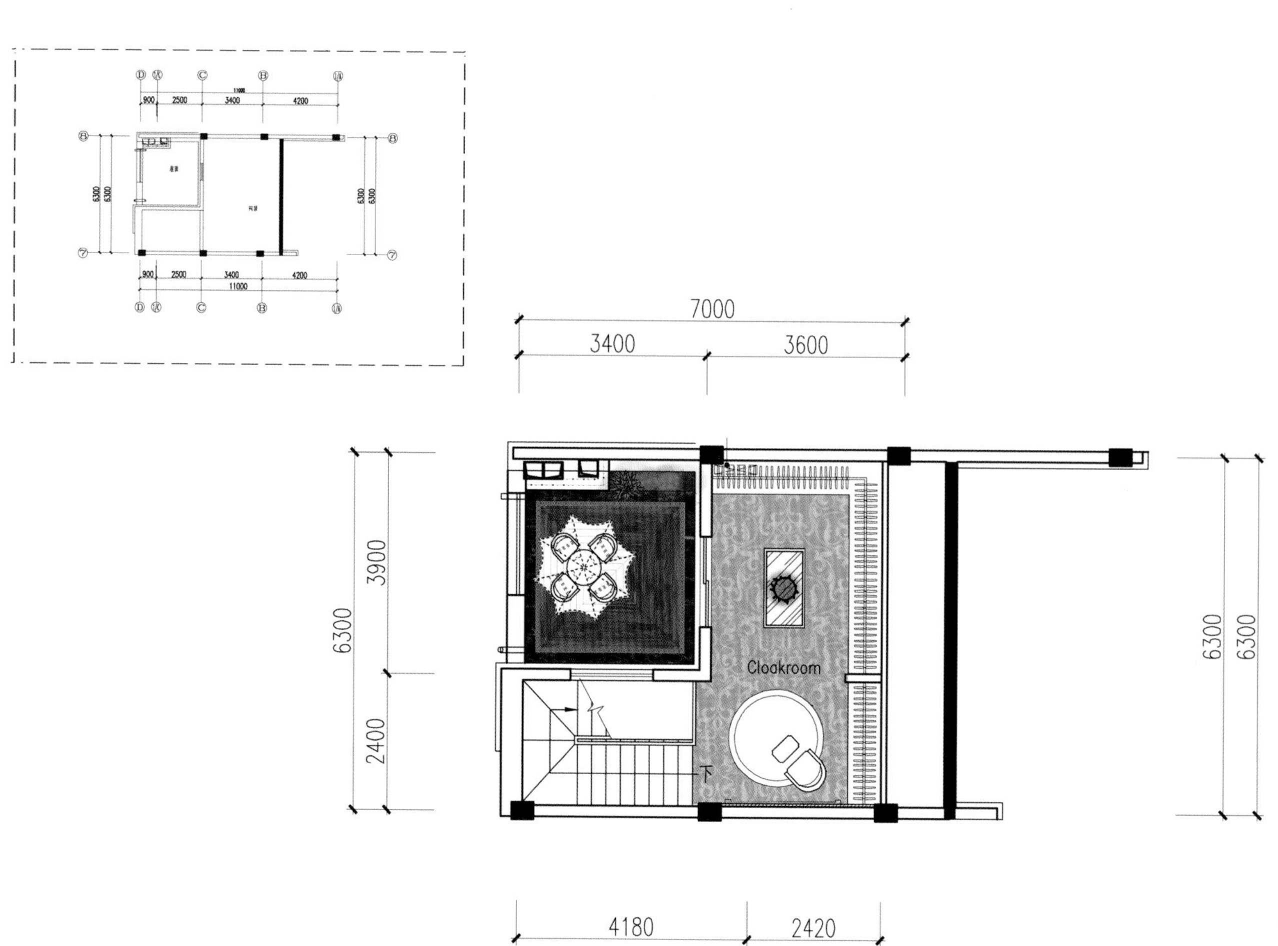

Roof Floor Plan
顶层平面图

Shunmai Villa Showroom, House Type 5

顺迈别墅样板间 – 五号户型

Designer: Wang Junqin
Interior Design: Wisdom Design Co., Ltd
Location: Hulan District, Harbin City, Heilongjiang, China
Area: 515 m²

设计师：王俊钦
室内设计：睿智汇设计
项目地点：中国黑龙江省哈尔滨市呼兰区
面积：515 m²

Keywords 关键词

Modern Breath 现代气息

Rococo Style 洛可可风格

Elegant & Gorgeous 优雅华丽

Furnishings/Materials 软装 / 材料

Black Mirror Steel, Antique Silver Foil, Stone, Grey Pole Rock Veneer, Film/Wired Glass, Leather

黑镜钢、仿古银箔、石材、灰色极岩木饰面、贴膜 / 夹丝玻璃、皮革

The house is made up of a basement and three ground floors. The first floor is functionally divided into hallway, bathroom, living room, opening kitchen and dinning area. The latter three areas are in wide-open layout and gracious with obstacle-free and spacious feelings. The extending upwards and curved surface ceiling of the living room stretches the visual height and the sunlike chandelier with strong modern breath lights up each corner of the space, which is low-key and energetic and enhances the fun of life. The dinning area upgrades the gorgeous temperament of the house. The wall and the ceiling with rich content are inspired from the petal and look as if can hold the whole world. The dining area also uses "sun" chandelier with strong modern breath and it polishes the charm of the petal.

The main function of the second floor is parents' room and children's room. The design of the parents' room is stable without losing vitality. The designer skillfully combines the changing room and the reading space with push-pull type partition that can casually transform the space. The hallway is designed in the MINI BAR mode, just like a small beverage bar of a hotel guestroom. The green bedside background with rich vitality is to meet the psychological needs of the aged and the sense of relaxing family life. The ceiling is decorated in flower totems and the lights are mainly gentle, linear and conceal in company of tube lamps.

The third floor is the main bedroom and the guestroom. The main bedroom is divided into several function areas such as changing room, bathroom, working area and leisure terrace. Flower is the design theme, and the flower patterns on the ceiling in company of the bed valance on the wall make people living here feel pleasant, comfortable, gracious and romantic. The master's bedroom is designed boldly

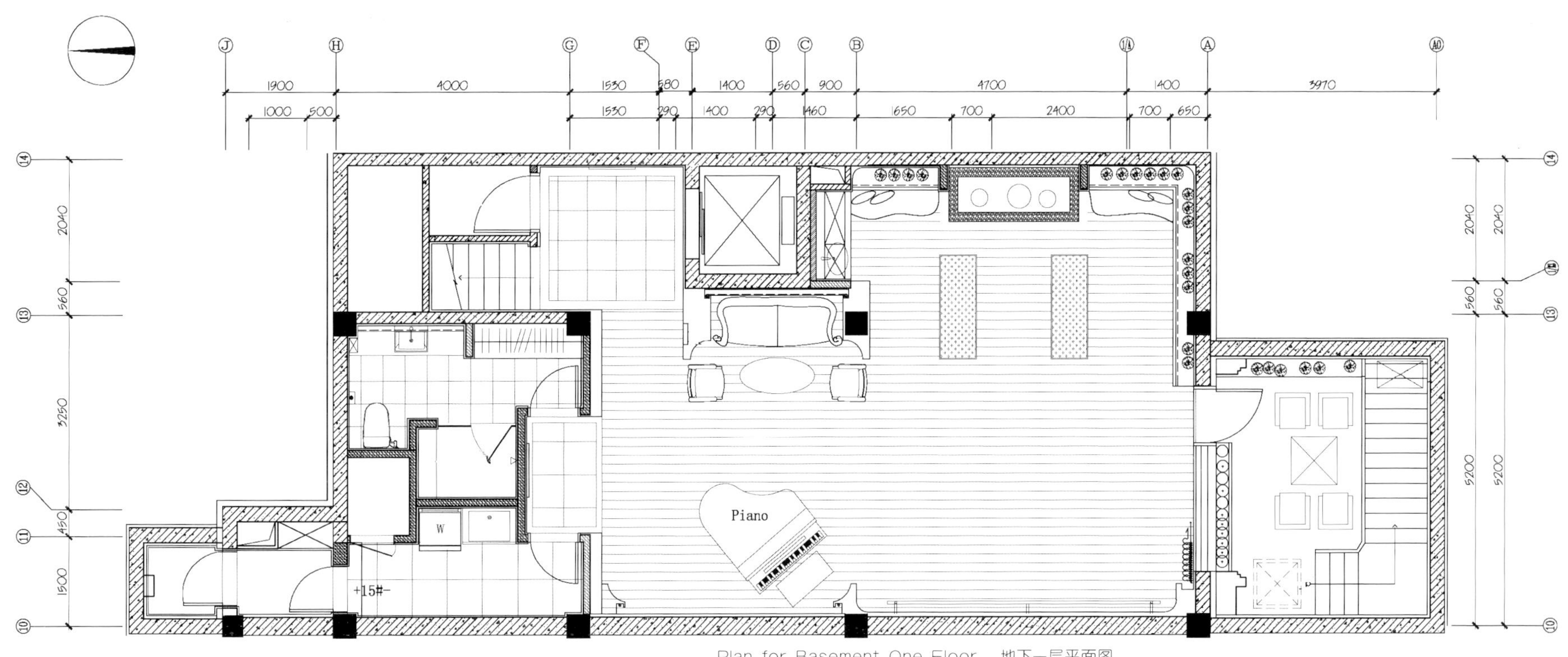

Plan for Basement One Floor　地下一层平面图

and innovatively with half-opened bathroom and the wide-opened bathtub. And the closestool and the shower are set on two sides and separated by glass partition. The leisure terrace of the master's bedroom can be planted with flowers which will provide great views and space for gathering.

The basement floor is mainly used for recreation, and it is equipped with areas that the hostess prefers, such as dancing area, yoga area, piano area, tea room and storeroom. The floor uses solid wood with good foot feelings, and the art painting of impressionism is also striking.

项目分为地下一层，地上三层。一楼功能布局分别有玄关、卫生间、客厅、开放式厨房及用餐区，其中，客厅、用餐区、厨房之间做了完全开放的布局设计，这种无障碍的开阔感受更显大气。客厅吊顶设计利用了向上延伸的弧形面，以此拉伸了视觉高度，配合现代气息浓郁的“太阳”吊灯，宛如挂在空中的太阳，照亮这空间的每一个角度，低调又蕴含能量，增加了生活趣味。用餐区将项目的华贵气质提升，墙面与顶面的灵感源于花瓣，内容丰富像是容纳了整个世界，用餐区同样选用了富有现代感的“太阳”吊灯，更加增添了花瓣的魅力。

二楼主要功能为双亲房及儿童房，双亲房设计沉稳又不失活力，设计师将更衣室与读书空间巧妙结合，推拉式隔断可以任意转换空间属性。同样为了完善功能性，设计师将入门处设计成MINI BAR形式，类似酒店客房中的小型饮品吧。床头背景选用了富有生机感的绿色，来迎合年迈老人需要的心理状态和轻松家居感受。天花选用花卉图腾装饰，灯光主要采用线性隐藏式灯光和筒灯的配合，光线柔和。

三楼是主人卧室和客卧，主卧分别有更衣室、卫生间、工作区、休闲露台几个功能布局。设计主题为花朵，顶面的花朵图案配合墙面的床幔，让生活在这里的人感到惬意和舒适，高雅而浪漫。主卧布局大胆创意，将卫生间功能半开放，其中浴盆完全开放，而马桶与淋浴间均采用玻璃隔断式放置于两侧，在开放的同时保证了基本的功能需要。主卧在建筑布局中设有休闲露台，可以栽培花卉植物，好友小聚的同时可以观赏园中景色。

地下一层主要以休闲功能为主，配有女主人偏爱的舞蹈区、瑜伽区、钢琴区、茶水间和部分储物空间等。地面选用脚感较好的实木地板，墙上的印象派艺术画亦引人注目。

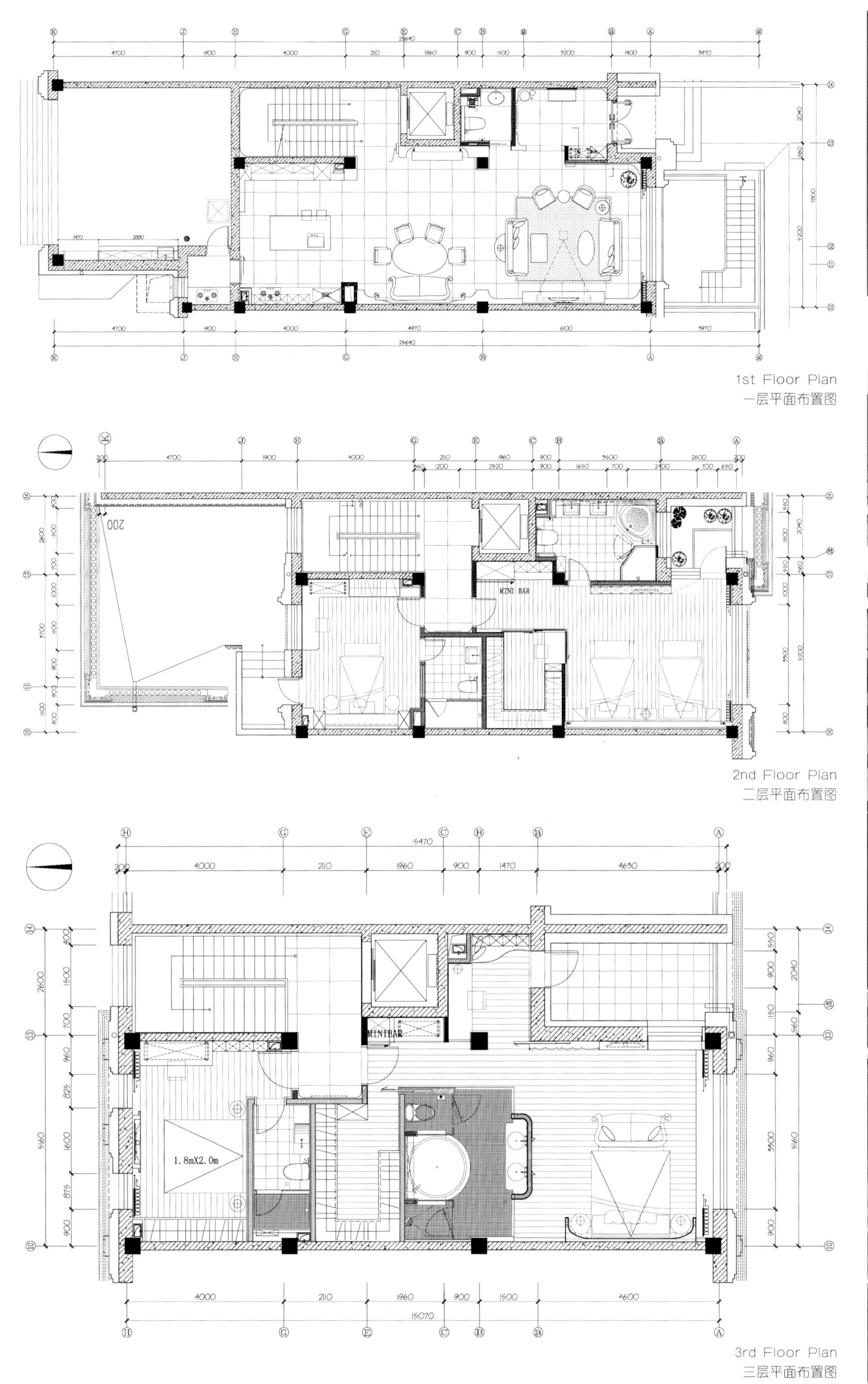

1st Floor Plan
一层平面布置图

2nd Floor Plan
二层平面布置图

3rd Floor Plan
三层平面布置图

BEAUTIFUL YOUTH